Dahlem Workshop Reports
Life Sciences Research Report 41

The goal of this Dahlem Workshop is:
to connect structural information
on humic substances to their
biogenesis and environmental role

Life Sciences Research Reports

Series Editor: Silke Bernhard

Held and published on behalf of the
Stifterverband für die Deutsche Wissenschaft

Sponsored by:
Senat der Stadt Berlin
Stifterverband für die Deutsche Wissenschaft

Humic Substances and Their Role in the Environment

F.H. Frimmel and R.F. Christman, Editors

Report of the Dahlem Workshop on
Humic Substances and Their Role
in the Environment
Berlin 1987, March 29 – April 3

Rapporteurs: J.M. Bracewell, J.R. Ertel,
H. Horth, E.M. Thurman

Program Advisory Committee:
F.H. Frimmel and R.F. Christman, Chairpersons
E.T. Gjessing, P.G. Hatcher, J.I. Hedges,
W. Ziechmann

A Wiley – Interscience Publication

John Wiley & Sons 1988
Chichester · New York · Brisbane · Toronto · Singapore

Copy Editors: J. Lupp, J. Lambertz, K. Klotzle

Photographs: E.P. Thonke

With 4 photographs, 62 figures, and 11 tables

British Library Cataloguing in Publication Data

Dahlem Workshop on Humic Substances and
   Their Role in the Environment (*1987 : Berlin*)
   Humic substances and their role in the environment:
   report of the Dahlem Workshop on Humic Substances
   and Their Role in the Environment, Berlin 1987, March 29–
   April 3.——(Dahlem workshop reports. Life Sciences
   research report; 41).
1.   Humus
   I.   Title II. Frimmel, F. H.   III. Christman, Russell F.
   IV. Series
   631.4'17   S592.8

ISBN 0 471 91817 2

Typeset by Photo·graphics, Honiton, Devon.
Printed and bound in Great Britain by the Bath Press Ltd., Bath, Avon.

Table of Contents

Table of Contents

The Dahlem Konferenzen

Founders
Recognizing the need for more effective communication between scientists, the Stifterverband für die Deutsche Wissenschaft*, in cooperation with the Deutsche Forschungsgemeinschaft**, founded Dahlem Konferenzen in 1974. The project is financed by the founders and the Senate of the City of Berlin.

Name
Dahlem Konferenzen was named after the district of Berlin called *Dahlem*, which has a long-standing tradition and reputation in the sciences and arts.

Aim
The task of Dahlem Konferenzen is to promote international, interdisciplinary exchange of scientific information and ideas, to stimulate international cooperation in research, and to develop and test new models conducive to more effective communication between scientists.

The Concept
The increasing orientation towards interdisciplinary approaches in scientific research demands that specialists in one field understand the needs and problems of related fields. Therefore, Dahlem Konferenzen has organized workshops, mainly in the Life Sciences and the fields of Physical, Chemical, and Earth Sciences, of an interdisciplinary nature.

Dahlem Workshops provide a unique opportunity for posing the right questions to colleagues from different disciplines who are encouraged to state what they do not know rather than what they do know. The aim is not to solve problems nor to reach a consensus of opinion, the aim is to define and discuss priorities and to indicate directions for further research.

The Donors Association for the Promotion of Sciences and Humanities, a foundation created in 1921 in Berlin and supported by German trade and industry to fund basic research in the sciences.
**German Science Foundation.

Topics
The topics are of contemporary international interest, timely, interdisciplinary in nature, and problem oriented. Dahlem Konferenzen approaches internationally recognized scientists to suggest topics fulfilling these criteria. Once a year, the topic suggestions are submitted to a scientific board for approval.

Program Advisory Committee
A special Program Advisory Committee is formed for each workshop. It is composed of 6-7 scientists representing the various scientific disciplines involved. They meet approximately one year before the workshop to decide on the scientific program and define the workshop goal, select topics for the discussion groups, formulate titles for background papers, select participants, and assign them their specific tasks. Participants are invited according to international scientific reputation alone. Exception is made for younger German scientists. Invitations are not transferable.

Dahlem Workshop Model
Since no type of scientific meeting proved effective enough, Dahlem Konferenzen had to create its own concept. This concept has been tested and varied over the years. It is internationally recognized as the *Dahlem Workshop Model*. Four workshops per year are organized according to this model. It provides the framework for the utmost possible interdisciplinary communication and cooperation between scientists in a period of 4 1/2 days.

At Dahlem Workshops 48 participants work in four interdisciplinary discussion groups. Lectures are not given. Instead, selected participants write background papers providing a review of the field rather than a report on individual work. These papers, reviewed by selected participants, serve as the basis for discussion and are circulated to all participants before the meeting with the request to formulate written questions and comments to them. During the workshop, each of the four groups prepares reports reflecting their insights gained through the discussion. They also provide suggestions for future research needs.

Publication
The group reports written during the workshop together with the revised background papers are published in book form as the Dahlem Workshop Reports. They are edited by the editor(s) and the Dahlem Konferenzen staff. The reports are multidisciplinary surveys by the most internationally distinguished scientists and are based on discussions of advanced new concepts, techniques, and models. Each report also reviews areas of priority interest and indicates directions for future research on a given topic.

The Dahlem Workshop Reports are published in two series:
1) Life Sciences Research Reports (LS), and
2) Physical, Chemical, and Earth Sciences Research Reports (PC).

Director
Silke Bernhard, Dr. med., Dr. phil. II h.c.

Address
Dahlem Konferenzen
Wallotstrasse 19
D-1000 Berlin (West) 33

Tel.: (030) 891 5067

THE DAHLEM WORKSHOP MODEL

MONDAY	TUESDAY	WEDNESDAY	THURSDAY	FRIDAY
A. Opening (P) B. Introduction (P) C. Selection of Problems for the Group Agendas (S) ① ② ③ ④	① ②	① ③	F. Report Session ① ② ③ ④	G. Distribution of the Reports H. Reading Time I. Discussion of the Group Reports (P)
D. Presentation of Group Agendas (P) E. Group Discussions (S) ① ②	③ ④	② ④		J. Groups Meet to Revise their Reports (S) ① ② ③ ④

Key: (P) = Plenary Session;
      (S) = Simultaneous Sessions;
      ◯ = one discussion group

Explanation of the Dahlem Workshop Model

A. Opening
   Background information is given about Dahlem Konferenzen and the Dahlem Workshop Model.
B. Introduction
   The goal and the scientific aspects of the workshop are explained.
C. Selection of Problems for the Group Agenda
   Each participant is requested to define priority problems of his choice to be discussed within the framework of the workshop goal and his discussion group topic. Each group discusses these suggestions and compiles an agenda of these problems for their discussions.
D. Presentation of the Group Agenda
   The agenda for each group is presented by the moderator. A plenary discussion follows to finalize these agendas.
E. Group Discussions
   Two groups start their discussions simultaneously. Participants not assigned to either of these two groups attend discussions on topics of their choice. The groups then change roles as indicated on the chart.
F. Report Session
   The rapporteurs discuss the contents of their reports with their group members and write their reports, which are then typed and duplicated.
G. Distribution of Group Reports
   The four group reports are distributed to all participants.
H. Reading Time
   Participants read these group reports and formulate written questions/comments.
I. Discussion of Group Reports
   Each rapporteur summarizes the highlights, controversies, and open problems of his group. A plenary discussion follows.
J. Groups Meet to Revise their Reports
   The groups meet to decide which of the comments and issues raised during the plenary discussion should be included in the final report.

Humic Substances and Their Role in the Environment
eds. F.H. Frimmel and R.F. Christman, pp. 1–2
John Wiley & Sons Limited
© S. Bernhard, Dahlem Konferenzen, 1988.

Introduction

F.H. Frimmel* and R.F. Christman**

*Engler-Bunte Institut der Universität Karlsruhe
7500 Karlsruhe 1, F.R. Germany
**Department of Environmental Sciences and Engineering
University of North Carolina
Chapel Hill, NC 27514, U.S.A.*

The need for communication between scientific disciplines concerned with common problems increases with scientific advancement, particularly when the problem is in the context of the natural environment and truly global processes are involved. Humic acid chemistry has been the purview of soil and agricultural scientists since the 18th century, and much of what is known today about the structure and properties of organic materials in terrestrial soil derives from this scientific area. In the last two decades, advances in analytical chemistry and the emergence of scientific interest in global organic geochemistry have led to the recognition that humic-like substances are ubiquitous in the earth's surficial water bodies and their sediments, as well as in many underground waters. Moreover, the presence of these complex natural substances has been found to markedly affect the productivity of plant systems, to influence the distribution of many other organic and inorganic chemical species in the environment, and to produce toxic by-products when exposed to water disinfectants containing halogen.

An important scientific communication need existed, therefore, between elements of the scientific community interested in the structure and properties of humic substances in the terrigenous and aquatic environments. The need found fertile ground at Dahlem Konferenzen, which invited a small Workshop Steering Committee to Berlin in February of 1986 to determine the principal features of the communication need and to identify a number of international scientists who would be willing to discuss the interdisciplinary

1

research frontiers of the chemistry of humic substances and their role in environmental processes.

The Workshop Steering Committee recognized that when natural substances are not fully defined in a chemical sense, precision, clarity, and consistency in the use of operationally defined terms are vitally important for scientific progress. The subject of the isolation of soil and aquatic humic substances became, therefore, the focus of the first of the four Workshop Discussion Groups. A second Discussion Group was assigned the fascinating task of exploring the pathways of humic substance formation, possible natural precursors, and the influence of the environment upon these.

Unlike many other natural organic products, humic substances are apparently devoid of any regularity in chemical structure. Chemical and enzymatic degradation approaches, which have been so fruitful with many natural biopolymers, are beset with enormous analytical problems when applied to humic substances. Have degradative techniques masked any regularity that might exist through outright destruction or the formation of intermediates so active that new polymers are formed? Are there important natural polymers that have been overlooked in the search for humic substance structural models? These important problems were given to the third Discussion Group. Finally, analysis of the important environmental reactions and processes controlled or affected by humic substances was deemed important enough to assign to our final Discussion Group.

A unique feature of the Dahlem Workshops provides for the Discussion Groups to determine their own agendas on the first day of the Workshop and to post a schedule of the times during which each item will be discussed. Since only two groups meet simultaneously, all Workshop participants have an opportunity to contribute to the discussion of all Groups. This is a remarkably effective process, owing to the fact that the feedback generated continuously modifies the activities of each Group. The net result is a more cohesive discussion of the four Workshop discussion areas, and for this reason, the final reports of each of the four Discussion Groups are published in this volume. The Dahlem Konferenzen organizers encourage the focusing of scientific discussion at the frontiers of research in the field and do not require that either the discussions or the published reports give the usual attention to comprehensive coverage of work in the field.

We believe that the quality of the product will be evident to any scientist interested in the earth's humic substances who chooses to read the material included in this volume. We cannot fully capture here the sense of cooperation, excitement, and stimulation that was evident at the Workshop in Berlin in the Spring of 1987. To those unable to participate, and to the devoted and hard working Staff of the Dahlem Konferenzen, we respectfully dedicate this volume.

Humic Substances and Their Role in the Environment
eds. F.H. Frimmel and R.F. Christman, pp. 3–14
John Wiley & Sons Limited
© S. Bernhard, Dahlem Konferenzen, 1988.

Isolation of Humic Substances from Soils and Sediments

J.W. Parsons

Department of Soil Science
University of Aberdeen
Aberdeen AB9 2UE, Scotland

Abstract. A continuing problem in studies on humic substances in soils and sediments has been our inability to compare results of chemical and physical analyses due to the variability of the isolation methods used. The International Humic Substances Society has introduced a standard procedure based on a four hour extraction with 0.1 mol/l NaOH at ambient temperature and under N_2. This paper reviews some of the factors influencing the extraction procedure and raises questions about its validity and whether it can be improved. The questions are:

1) In the case of soil samples should an initial flotation procedure be included to remove plant remains?
2) Should we follow the example set in the extraction of sediments and introduce a pretreatment with organic solvents to remove lipids and alkanes from soils?
3) What is the status of carbohydrate and amino compounds? Are they part of the humic structure or should they be separated if necessary by acid hydrolysis?
4) In what form does the fulvic acid fraction occur? Does it make up a separate identifiable humic fraction or is it produced as a result of the extraction procedure?
5) Can we develop a procedure which will extract the fulvic acid fraction components before extracting humic acid fractions?

These and other questions can form the basis of our discussions.

INTRODUCTION

The organic components of soils and sediments play a vital role in terrestrial and aquatic ecosystems, consequently their chemical composition has intrigued biologists and chemists for a very long time. Chemical analyses of untreated samples have provided limited information on their structure but it was recognized from very early on that an extraction step to separate organic from the inorganic components must be achieved as a preliminary stage before detailed chemical analyses can be embarked upon. The choice of extractant is limited by the complex heterogeneity of the material and the rigorous demands placed upon the extractant. Stevenson (1982) listed the important characteristics of an ideal extractant as follows:

1) The method leads to the isolation of unaltered material.
2) The extracted humic substances are free of inorganic contaminants such as clay and polyvalent cations.
3) Extraction is complete, thereby ensuring representative fractions from the entire molecular weight range.
4) The method is universally applicable to all soils.

Do the solvents available to us meet all these criteria? The answer to that question must be no because of the solubility limits imposed by chemical composition, molecular weight, and charge characteristics. However, it is important that we fully understand what is happening during extraction to minimize the possibility of producing artifacts, and to choose standard procedures which ensure that we are working with comparable products.

Hayes (1985) has provided an excellent review of the physicochemical basis of extraction and, though inevitably some of the same ground will be covered here, I will also attempt to examine the topic from a slightly different viewpoint and raise questions which are sometimes neglected. This review will deal mainly with isolation of humic substances from soils because of the much larger literature on that subject compared to work on sediments.

WHAT IS THE COMPOSITION OF THE STARTING MATERIAL?

In raising this question I have no intention of reviewing the range of theories available to us on chemical structure but rather to consider the composition of the sample chosen for extraction. A cursory look at air-dried soil highlights the problems. The dark brown to black coloration of the aggregates provides visual evidence for the presence of humic compounds adsorbed on mineral surfaces but, depending on soil type, there will be variable proportions of plant, animal, and microbial remains. Humic substances have been defined as naturally occurring biogenic, heterogeneous organic substances that can generally be characterized as being yellow to black in color, of high molecular weight, and refractory. In addition, they are the products of the humification process and in theory we can talk about humified or "true" humic substances and nonhumified materials, but this division is purely arbitrary and can rarely be achieved in practice.

German soil chemists in the 1920s attempted to resolve the problem by treating soils with acetyl bromide or acetic anhydride to acetylate and dissolve nonhumified materials. There is no evidence to suggest they were successful and it is highly unlikely that such a selective solvent exists. Ford et al. (1969) applied a more logical approach to the problem by removing obvious plant and animal remains, which they called the "light fraction," by flotation on an inert organic liquid of suitable density [2 g cm^{-3}]. The light fraction accounted for between 9% and 70% of the total soil carbon, a significant fraction in many soils. Microbial cells are tightly held on surfaces and would not be removed by the flotation procedure except for those adsorbed on plant remains. In most extraction procedures obvious plant remains may be removed by hand, but rarely is flotation used as a pretreatment, and yet a significant amount of plant material will dissolve in subsequent alkaline extraction.

A flotation pretreatment of sediments is of little value but extraction of humic substances from sediments invariably follows an organic solvent extraction, generally ethanol-benzene or methanol-chloroform mixtures, to remove lipids and waxes. Soil chemists have advocated and indeed some have used organic solvent pretreatments of soils, but they are rarely used in standard procedures. However, Ogner and Schnitzer (1970) and Khan and Schnitzer (1972) have isolated alkanes and fatty acids by organic solvent extraction of humic and fulvic acids, although in both cases maximum release was only obtained after methylation of the humic materials. The humic fractions were obtained by alkaline extraction of soils involving no pretreatment, except a dilute acid extraction in one case, which raises the question, "did the complexes form during the extraction process?" Schnitzer and co-workers suggest that the fulvic acid appears to have the ability of converting hydrophobic compounds into water soluble complexes. If this is correct and the complexes form naturally in soils, this could be very important in studies on the fate of pesticides and other man-made hydrophobic chemicals in soils.

Other cellular components, such as carbohydrates, proteins, and mucopeptides, are not so easily removed from soils; in fact, acid hydrolysis at elevated temperatures is required to split what appear to be covalent bonds, linking them to humic fractions. Ishiwatari (1985) subtracted the estimated amounts of lipid, carbohydrate, and protein from total organic matter in sediments to provide an estimate of "true" humic components or "nonbiochemical" components. Riffaldi and Schnitzer (1973) suggest that "6 mol/l HCl hydrolysis appears to purify (soil) humic acids and so provides more homogeneous starting materials for subsequent analytical and structural investigations." Haworth (1971) was more emphatic in his belief that biochemical components are present as "impurities" and introduced a two-stage process to remove first carbohydrate and then protein material from humic acid to leave a resistant core.

Perhaps we should attempt to define in operational terms what true humic and fulvic acids are. Lipids, which account for a very small proportion of the humic-C, certainly do appear to be present as an admixture, although they are held by simple bonds. Whether carbohydrates and particularly protein or amino compounds which cannot be removed by physical means are part of the humic structure remains an important question which has not been adequately resolved and needs to be addressed.

EXTRACTION WITH ALKALI

The earliest reported reagent used for extracting humic substances was sodium hydroxide and to this day, with only minor modifications of the procedure, it remains the most widely used. Numerous other inorganic reagents and a number of organic solvents have been tried but none have proved so effective in terms of quantity extracted.

The factors which influence efficiency of extraction are concentration of reagent, ratio of extractant/soil, temperature of extraction, type of extractant, time of extraction, and pH. Most of these variables have been standardized and while extraction in alkaline solution at elevated temperatures (Evans 1959; Swift and Posner 1972) certainly dissolves more humic materials, extraction at 25°C is the preferred choice. This choice of temperature is based on our understanding of chemical reactions which may occur in alkaline media at elevated temperatures rather than careful experimental measurement. One of the few studies on the effect of NaOH extraction at 60°C, made by Cameron et al. (1972), showed that the humic acid isolated had a larger molecular size than that extracted with pyrophosphate.

Why are alkaline extractants so effective?

Apart from temperature the two variables which most influence alkaline extraction efficiency are time and pH and they appear to be interrelated. Bremner (1949) showed that, while time has little effect on extraction with neutral pyrophosphate, extraction of C and N in 0.5 mol/l NaOH increased with time over a period of 50 hours. This he suggested may be due to a slow depolymerization of humic molecules forming products soluble in alkali. Using gel-permeation chromatography, Cameron et al. (1972) were able to show that NaOH extracted humic acid has a larger molecular size than that extracted by pyrophosphate, indicating that hydrolysis does not appear to be a major problem over a 24 hour period, but extension of this period to 30 days leads to a significant decrease in molecular weight (Swift and Posner 1972).

In soil studies, 0.5 mol/l NaOH is the preferred strength of extractant because the concentration of Na$^+$ ions causes flocculation of clay particles. There is little to choose between 0.1 and 0.5 mol/l in terms of efficiency, and in work on sediments the weaker solution is generally used. However, pH is a critical factor in the dissolution of humic substances. The results, presented in Fig. 1, were obtained from an extraction in which the Na$^+$ ion concentration was maintained at 0.5 mol/l with NaCl and the pH was carefully controlled through a 20 hour extraction period (Parsons 1981).

Nitrogen rather than carbon was taken as a measure of humic material extracted for convenience of the analytical determination. The amount extracted increased with increasing pH; even between pH 13, a pH value which cannot be considered too reliable, and extracting with 0.5 mol/l NaOH there was a significant increase. Humic substances in soils and sediments occur as metal salts or they are adsorbed to mineral surfaces, and one reason for their solubilization in alkali is the conversion of the bound functional groups to anionic groups in the soluble sodium salt. Why such a high pH is

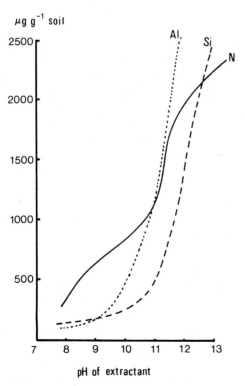

Fig. 1—Extraction of N, Al, and Si from soil.

Table 1 Liberation of ammonia from a suspension of soil in 0.5 mol/l NaOH over a 24 hour period.

Time	NH_3-N released as % of soil-N
2	0.2
4	0.4
6	0.7
8	1.1
10	1.5
12	1.9
24	3.7

necessary is not immediately obvious but bonds linking humic substances to other inorganic or even organic components will be severed. It is significant that solution of both aluminum and silicon follow a similar pattern to that of extraction of nitrogen (Fig. 1). It is interesting to note that Gascho and Stevenson (1968) failed to obtain the relationship with pH for a soil which had received a sequential pretreatment with 0.3 mol/l HF and 0.1 mol/l sodium pyrophosphate.

The use of alkaline extractants has been criticized for many reasons. It has been known for some time that humic substances take up oxygen in alkaline solution; Swift and Posner (1972) have shown that uptake increases as pH increases. Over an extraction period of 30 days the contents of keto and carboxyl groups increased and acid-insoluble humic acids were converted to acid-soluble material. Standard extraction procedures are now conducted in an atmosphere of nitrogen.

Amide and some amino groups are labile in alkaline solution. Ammonia-free air blown through a suspension of soil in 0.5 mol/l NaOH liberated small amounts of ammonia (see Table 1).

The amount of ammonia released, although small, was the same whether air or N_2 gas was used and is of the same order as that determined as amide-N in soils. There are two aspects of this problem: (a) the loss of labile N groups and (b) the possibility of the ammonia liberated reacting with other components, particularly quinones, at high pH during extraction to produce artifacts.

An intriguing question posed by the fractionation of humic extracts from soils is "in what form do fulvic acids occur in soils?" Berzelius, in the early 19th century, coined the name *crenic acid* for organic material extracted from soils by water, and a century later Odén introduced the term *fulvic acid* to describe the water soluble organic material expressed from peat. These terms, used in their original form, describe the water soluble material studied by chemists at the present time in aquatic and marine systems. The fulvic acid fraction is now defined quite differently and is restricted to the material dissolved from soils in alkali which remains soluble on acidification. Unfortunately the terms *humic* and *fulvic acids*, orginally chosen to describe fractions isolated from soils, have been purloined by those working with freshwater and marine organic substances, which may be very different both in origin and in chemical composition.

Schnitzer chose to do much of his earlier work on fulvic acid fractions extracted directly from the illuvial (B) horizon of a Spodosol with dilute hydrochloric acid. However, for the majority of soils, dilute acid extraction fails to dissolve much fulvic material and yet once an alkaline extract has been acidified a significant proportion remains in acid solution. Why should organic material previously insoluble in acid become acid soluble?

The ratios of humic to fulvic acid fractions obtained by sodium hydroxide extraction from a range of soils, quoted by Kononova (1966), vary from 0.3 to 2.5. This suggests that big differences occur between soil types but at the same time other evidence points to a relationship between the ratio of the two fractions and the concentration and pH of the extractant. Evans (1959) has shown that as pH of the extractant is increased from 7 to 11.7 the proportion of fulvic acid decreases from 81% to 56% of the extracted material. In other words, the ratio of humic to fulvic acids increased with increasing pH. Gascho and Stevenson (1968) obtained the opposite result with a soil which had received an HF pretreatment. They also showed that increasing the concentration of sodium pyrophosphate from 0.01 to 0.25 mol/l caused a decrease in the humic to fulvic ratio. Evans (1959) suggests that the fulvic acid fraction dissolved at near neutral pH and as the pH is increased more humic acid fraction was brought into solution, whereas the results of Gascho and Stevenson (1968) indicate that increasing pH and salt concentration may be severing bonds, such as ester bonds, between humic and fulvic acids.

The fulvic acid fraction, by the very nature of the isolation procedure, contains high concentrations of salt and heavy metal ions and therefore tends to be neglected and often thrown down the drain. However, Gregor and Powell (1986) have recently introduced a procedure using $H_2P_2O_7^{2-}$ at pH 2 to extract the fulvic acid fraction directly with minimal contamination from humic acid. They took great care not to raise the pH above 6.5 and not to leave the extracts to stand at a low pH; they were able to show a general increase in the molecular weight of the fulvic acid fraction maintained at pH 1.7 for a period of 7 days. Their results were obtained with B_h horizon samples from a gleyed Podzol, which is perhaps not a fair comparison with surface horizons, but they did obtain significant results from a peat sample. No data is provided to compare extraction either with dilute HC1 directly or with sodium hydroxide followed by acidification.

Adsorption of humic materials on silicate surfaces or in the interlayer space of clay minerals is often given as the explanation for the failure to achieve their complete extraction from soils even with strong alkali. This may be one contributing factor but it does not provide the complete answer because a pretreatment with dilute HF to destroy silicates, although it increases the efficiency of a subsequent alkaline extraction, still leaves a significant fraction, the humin, in the soil residue. Similarly, ball-milling the samples before extraction to expose new surfaces has little effect on subsequent alkaline extraction. Dilute HCl is often used as a pretreatment but it is only of value in the case of calcareous soils.

NEUTRAL AQUEOUS EXTRACTANTS

A wide range of aqueous solutions of sodium salts, such as pyrophosphate, citrate, fluoride, etc., which form complexes with heavy metal ions, have been used with variable success. They have the merit of reducing the possibility of producing artifacts in the neutral medium but they are considerably less effective than alkaline extractants. Efficiency of extraction is not greatly affected by time of extraction but Evans reported that maintaining the temperature, over a two-day period at 57°C, more than doubled the amount extracted with neutral pyrophosphate at 17°C. Pyrophosphate is the most widely used and most efficient of the neutral salts although it is not always clear whether pyrophosphate has been used at its natural pH (9.5) or adjusted to pH 7. Ion exchange and chelating resins in the Na form have been tried but the efficiency of extraction appears to be related to the pH of the suspension rather than the ability of the resin to take up metal ions. Extractions with distilled water or with defined sodium chloride solutions can be useful in studying the water soluble part of soils or sediments in relation to seepage water problems and to immobilization reactions in estuarine systems.

ORGANIC EXTRACTANTS

Organic extractants, as mentioned earlier, are used in both soil and sediment studies to remove a specific group of components as a pretreatment. Common organic solvents, such as ethanol or acetone, are ineffective on their own but addition of HCl to an aqueous acetone mixture (Porter 1967) does significantly increase the amount of humic material dissolved from soil and leonardite, presumably by breaking metal-organic bonds.

Hayes (1985) has provided a comprehensive review of the work done, both by his own group and that of others, on extraction with dissolved organic substances and organic solvents such as ethylenediaminotetraacetate, diaminoethane, N,N-dimethyl formamide, and dimethyl sulfoxide, either alone or in mixtures containing alkali, acid, or sodium pyrophosphate. Dimethyl sulfoxide, in combination with acid, is the most effective but recovery of the dissolved humic material is achieved by making the solution alkaline. Another serious disadvantage with nitrogen- or sulfur-containing solvents is the difficulty of completely removing the solvent from the final humic product.

I have used formic acid, a polar solvent, which in its anhydrous state possesses the characteristics of a good solvent for humic material (Parsons and Tinsley 1960). A short (15 min) extraction at 100°C extracts a fraction with a low N content but containing a high proportion of carbohydrate, which is easily recovered by precipitation with excess ether. The effectiveness of formic acid was greatest with soils low in clay and silt and subsequent work suggested that part of the humic material was actually adsorbed onto clay surfaces during extraction (Watson and Parsons 1974). Pretreatment of a clay soil with 1 mol/l HF (Parsons 1981) more than doubled the amount of C and N extracted in formic acid and at the same time increased the per cent of N in the extracted material and reduced the inorganic contamination of the product. The total amount of humic material extracted with formic acid following a 1 mol/l HF pretreatment ranged from 44% to 68%.

The use of anhydrous formic acid as an extractant overcomes the problems of high pH and oxidation but substitutes problems associated with an acidic medium. However, there is little or no evidence for hydrolysis, even at 100°C providing the medium is kept anhydrous.

RECOVERY OF ISOLATED MATERIAL

From the earliest studies on the alkaline extraction of organic components of soils, an acidification step has been included to precipitate the humic acid fraction, both as a purification and fractionation exercise. This step has become so ingrained in the methodology and in the operationally derived definitions of the fractions that it is widely applied to all aqueous and some organic extracts from soils. The crude humic acid fraction is contaminated with metal ions (in particular iron, aluminum, and silicon) and may be coprecipitated with both fine clay particles and "biochemical" components of plant, animal, and microbial origin.

Humic acid fraction

For studies of humic-metal complexes it would be advantageous to extract the complexes intact. Desirable as this may be, it is virtually impossible to attain because the metal ions associated with the functional groups are partially responsible for the insolubility of the polymers. Generally, removal of the metals is a declared objective of the isolation procedure and can be achieved by redissolving and reprecipitation, followed either by suspending the precipitate in dilute acid and then washing or by dialyzing against dilute acid (HCl and HF) followed by water. This step invariably leads to loss of small molecular weight material which should be combined with the fulvic acid fraction. It is essential to note that failure to completely remove mineral acid before drying can lead to erroneous results in subsequent estimations of functional groups.

Fine clay particles can be removed by redissolving the humic acid precipitate in alkali, adjusting to pH 7, and applying high speed centrifugation. Dialysis against dilute HF will also remove silicate material.

Removal of nonhumic material, even if it could be easily defined, is difficult in practice. Various fractionation methods have been used and many have introduced terminology which is gradually disappearing from the literature because of the narrow aim of the procedure and the limited possibility of interpretation of the operationally defined values. An example is *hymatomelanic acid* which is the alcohol soluble portion of the humic acid fraction. Scharpenseel and Krausse (1962) fractionated humic acid into *brown* and *gray humic acids* by saltation and obtained a concentration of nitrogen in the brown humic acid fraction. Biederbeck and Paul (1973) extracted a nitrogen-rich fraction from humic acid with phenol and, after fractionation on polyvinylpyrrolidone, they isolated an aliphatic nitrogen-rich component which they considered to be reversibly combined with humic acid by hydrogen bonding. Other components, such as amino acids, amino sugars, and nitrogen bases of nucleic acid origin, have to be separated by strong acid hydrolysis. Once again we have to ask whether compounds attached by covalent linkages are an integral part of the humic acid structure or not.

Fulvic acid fraction

This fraction is heavily contaminated with salt from the extraction procedure and with metal ions passing into the acidic solution. A charcoal/celite mixture and polyvinylpyrrolidone (Cheshire et al. 1974) and Amberlite XAD-8 resin (Anderson et al. 1987) have been used to separate polysaccharide and amino compounds from the fulvic acid fraction.

CONCLUSIONS

Despite the possible weaknesses of alkaline extraction, the procedure (providing certain precautions are observed) remains the most effective and widely used. In the absence of any obvious and exciting new solvents, tremendous progress has been made in the last few years by the introduction of a standardized extraction procedure based on sodium hydroxide by the International Humic Substances Society. There may still be some doubts about the advisability of using an alkaline medium, but in the absence of a superior reagent we can at least proceed to make measurements and analyses which will produce results that can be compared with others on a rational basis. Perhaps it is time to reappraise the extraction scheme to see if any further improvements can be made. Perhaps the simple answer would be to adopt a standard procedure using sodium hydroxide in which all the variables are controlled to minimize artifact production rather than to maximize yield of isolated material.

REFERENCES

Anderson, H.A.; Bick, W.; Hepburn, A.; and Stewart, M. 1987. Nitrogen in humic substances. In: Humic Substances II In Search of Structure, M.H.B. Hayes, P. MacCarthy, R.L. Malcolm and R.S. Swift, eds., Chap. 8. New York: John Wiley-Interscience, in press.

Biederbeck, V.O., and Paul, E.A. 1973. Fractionation of soil humate with phenolic solvents and purification of the nitrogen-rich portion with polyvinylpyrrolidone. *Soil Sci.* **115**: 357–366.

Bremner, J.M. 1949. Studies on soil organic matter. Part III. The extraction of soil carbon and nitrogen from soil. *J. Agr. Sci.* **39**: 280–282.

Cameron, R.S.; Thornton, B.K.; Swift, R.S.; and Posner, A.M. 1972. Molecular weight and shape of humic acid from sedimentation and diffusion measurements on fractional extracts. *J. Soil Sci.* **23**: 394–408.

Cheshire, M.V.; Greaves, M.P.; and Mundie, C.M. 1974. Decomposition of soil polysaccharide. *J. Soil Sci.* **25**: 483–498.

Evans, L.T. 1959. The use of chelating reagents and alkaline solutions in soil organic-matter extraction. *J. Soil Sci.* **10**: 110–118.

Ford, G.W.; Greenland, D.J.; and Oades, J.M. 1969. Separation of the light fraction from soils by ultrasonic dispersion in halogenated hydrocarbons containing a surfactant. *J. Soil Sci.* **20**: 291–296.

Gascho, G.J., and Stevenson, F.J. 1968. An improved method for extracting organic matter from soil. *Soil Sci. Soc. Am. Proc.* **32**: 117–119.

Gregor, J.E., and Powell, H.K.J. 1986. Acid pyrophosphate extraction of soil fulvic acids. *J. Soil Sci.* **37**: 577–585.

Haworth, R.D. 1971. The chemical nature of humic acid. *Soil Sci.* **111**: 71–79.

Hayes, M.H.B. 1985. Extraction of humic substances from soils. In: Humic Substances in Soil, Sediment, and Water. G.R. Aiken, D.M. McKnight, R.L. Wershaw, and P. MacCarthy, eds. New York: John Wiley-Interscience.

Ishiwatari, R. 1985. Geochemistry of humic substances in lake sediments. In: Humic Substances in Soil, Sediment, and Water. G.R. Aiken, D.M. McKnight, R.L. Wershaw, and P. MacCarthy, eds. New York: John Wiley-Interscience.

Khan, S.U., and Schnitzer, M. 1972. The retention of hydrophobic organic compounds by humic acid. *Geochim. Cosmochim. Acta* **36**: 745–754.

Kononova, M.M. 1966. Soil Organic Matter: Its Nature, Its Role in Soil Fertility, 2nd ed. Oxford: Pergamon.

Ogner, G., and Schnitzer, M. 1970. The occurrence of alkanes in fulvic acid, a soil humic fraction. *Geochim. Cosmochim. Acta* **34**: 921–928.

Parsons, J.W. 1981. Clay-organic nitrogen complexes in soils. Proc. of Colloque Humus-Azote, 53–59. P. Dutil, and F. Jacquin, eds. National Supérieure d'Agronomie et des Industries Alimentaires, Nancy, France.

Parsons, J.W., and Tinsley, J. 1960. Extraction of soil organic matter with anhydrous formic acid. *Soil Sci. Soc. Am. Proc.* **24**: 198–201.

Porter, L.K. 1967. Factors effecting the solubility and possible fractionation of organic colloids extracted from soil and leonardite with an acetone-water-hydrochloric acid solvent. *J. Agri. Food Chem.* **15**: 807–811.

Riffaldi, R., and Schnitzer, M. 1973. Effects of 6N HCl hydrolysis on the analytical characteristics and chemical structure of humic acids. *Soil Sci.* **115**: 349–356.

Scharpenseel, H.W., and Krausse, R. 1962. Investigation of amino acids in various organic sediments especially gray- and brown-humic acid fractions of different soil types (including C^{14}-labelled humic acids). *Zeitschrift für Pflanzenernährung Düngung Bodenkunde* **96**: 11–34.

Stevenson, F.J. 1982. Humus Chemistry. New York: John Wiley-Interscience.

Swift, R.S., and Posner, A.M. 1972. The distribution and extraction of soil nitrogen as a function of soil particle size. *Soil Biol. Biochem.* **4**: 181–186.

Watson, J.R., and Parsons, J.W. 1974. Studies of soil organo-mineral fractions. II. Extraction and characterisation of organic nitrogen compounds. *J. Soil Sci.* **25**: 9–15.

Humic Substances and Their Role in the Environment
eds. F.H. Frimmel and R.F. Christman, pp. 15–28
John Wiley & Sons Limited
© S. Bernhard, Dahlem Konferenzen, 1988.

A Critical Evaluation of the Use of Macroporous Resins for the Isolation of Aquatic Humic Substances

G.R. Aiken

U.S. Geological Survey,
Water Resources Division
Arvada, CO 80002, U.S.A.

Abstract.   Column chromatographic methods using either the nonionic XAD resins or the anionic resin, Duolite A-7, are currently the most commonly used methods for isolating humic substances from water. While these methods are very efficient, it is important to recognize their limitations. Presently, there is no universally accepted method of isolating aquatic humic substances and no consensus as to the operational definition of these materials. Consequently, direct comparison between different samples is difficult, and a portion of the hydrophilic acid fraction, which is chemically very similar to aquatic humic substances, is often ignored. In addition, the isolation process results in chemical alteration of the isolated material. While a certain amount of alteration is unavoidable in removing this material from the environment, other potential sources of alteration (including ester hydrolysis, alcoholysis reactions, and contamination by resin bleed) are associated with the method of extraction. Understanding the principles involved in sample processing is important in designing systems that will minimize sample alteration or contamination.

INTRODUCTION

The study of the chemistry and environmental significance of aquatic humic substances is greatly dependent on the limitations imposed on the sample by the methods used to isolate this material. Indeed, these materials are operationally defined based upon extraction procedures. In the environment, humic substances are present in solution at relatively low concentrations and

Table 1 Dissolved organic carbon fractionation data (mgC/L) for select freshwater systems in the United States.

Sample	DOC	Hydrophobics			Hydrophilics		
		Acids	Bases	Neutrals	Acids	Bases	Neutrals
Ohio River	3.7	1.2	0.0	0.9	1.3	0.3	0.0
Missouri River	3.4	0.7	0.8	0.0	1.7	0.3	0.0
Suwannee River	38.2	16.0	0.2	1.0	19.2	1.4	0.6

interact with other dissolved species, both organic and inorganic. As part of the process of isolating these materials, many interactions are disrupted and changes in the chemical structure of the compounds themselves may result. It is very important, therefore, that the methods used for extraction and isolation be evaluated critically and that the results of subsequent analyses on the extracted material be interpreted with any drawbacks of the extraction method in mind.

Organic matter in water can be divided into dissolved and particulate organic carbon. Since no natural cutoff exists between these two fractions the distinction is operational, e.g., filtration through a 0.45 micron filter has been established arbitrarily as the standard procedure for separating the dissolved and particulate components (Danielsson 1982). It is important to note that research on humic substances in water has been concerned almost exclusively with the dissolved fraction.

Dissolved organic carbon can be divided into six fractions: hydrophobic acids, bases, and neutrals; and hydrophilic acids, bases, and neutrals (Leenheer 1981). Dissolved organic carbon (DOC) data from a variety of freshwater systems are presented in Table 1. Humic substances are the major component of the hydrophobic acid fraction. These compounds range in concentration from 20 µg/l in groundwater to over 30 mg/l in surface water (Thurman and Malcolm 1981). Before their chemical properties can be defined thoroughly, aquatic humic substances must be isolated from other organic compounds and inorganic species, such as metal ions and dissolved silica. In the last 25 years, a number of methods have been developed to isolate efficiently and concentrate humic substances from water. A review of the available methods has appeared in the literature recently (Aiken 1985).

Currently, the most widely used methods for isolating humic substances from water are column chromatographic methods using the nonionic XAD resins or the anion-exchange resin, Duolite A-7. Advantages of these methods include easy handling of large volumes of water, high concentration factors for isolated solutes, fractionation of dissolved organic solutes according to sorption characteristics, and regeneration of sorbent. The purpose of this paper is to critically evaluate these methods for isolating aquatic humic substances from water and to identify potential sources of sample alteration resulting from use of these methods.

OVERVIEW OF METHODS

Amberlite XAD resins have been used extensively as adsorbents for organic solutes from water during the last 15 years, and sorption of organic solutes on this resin series has been studied extensively (Thurman et al. 1978). These resins have large surface areas and, in general, sorption of organic acids is determined by a solute's aqueous solubility and the solution pH. Theoretically, hydrophobic organic acids in water can be isolated from the remainder of the organic solutes present, including the hydrophilic organic acids, by properly selecting the influent pH and the volume of sample to be processed for a given column size (Aiken 1987). At low pH, weak acids are protonated and adsorbed on the resin; at high pH, weak acids are ionized and desorption is favored. Differences in resin pore size, surface area, and chemical composition result in different capacity factors for the same solute on each resin.

Comparisons of commonly used XAD resins have been published for the isolation of both fulvic acid (Aiken et al. 1979) and humic acid (Cheng 1977) from freshwater and seawater (Mantoura and Riley 1975). These resins differ in pore size, surface area, polymer composition, and polarity. The styrene-divinylbenzene resins (XAD-1, XAD-2, and XAD-4) were found to be more difficult to elute than the more hydrophilic acrylic-ester resins (XAD-7 and XAD-8) for terrestrially derived humic substances. This is due to hydrophobic interactions and possible π-π interactions with the aromatic resin matrix of the styrene-divinylbenzene resins. In addition, the acrylic-ester resins wet more easily and absorb more water than the styrene-divinylbenzene resins. This results in faster sorption kinetics on the acrylic-ester resins. The acrylic-ester resins have higher capacities for terrestrially derived fulvic acid and are eluted more easily than the styrene-divinylbenzene resins. Of the various XAD resins available, the most commonly used for the isolation of humic substances from water are XAD-2 (styrene-divinylbenzene) and XAD-8 (acrylic-ester).

Thurman and Malcolm (1981) have published a procedure using XAD-8 that is commonly used for isolating humic substances from freshwater. According to this method, the sample to be processed is initially filtered through a 0.45 micron silver filter. After acidification to pH 2 with HCl, the sample is passed through a column of XAD-8 resin. During this step the humic substances adsorb to the resin. The humic substances are then back eluted from the column with NaOH, hydrogen saturated, and lyophilized. Similar procedures are followed when XAD-2 is used (Mantoura and Riley 1975)

Use of anion-exchange resins for concentrating organic acids from water was reported as early as 1964 by Packham (1964). The criteria necessary for an efficient anion sorbent of aquatic humic substances are weak-base functional groups, a macroporous structure, and a hydrophilic matrix that is negatively charged at pH 10 (Abrams 1975). Duolite A-7 is a macroporous, weak-base resin, with a phenol-formaldehyde matrix. This resin has a high capacity for anionic organic solutes and excellent elution characteristics when the loading is limited to one-half to two-thirds of the resin capacity. All the organic acids and inorganic anions present in a water sample are concentrated on the A-7 resin. The inorganic anions are coeluted with the organic acids and are included in the column eluate. The humic substances can be desalted and separated from the hydrophilic acid fraction by adjusting the A-7 eluate to pH 2, and passing the solution through a column of XAD-8. XAD resins are suitable for both freshwater and seawater samples although methods utilizing ion exchange are not suitable for isolating humic substances from waters of high ionic strength such as seawater.

A method for isolating humic substances using the A-7 resin has been published by Leenheer and Noyes (1984). As shown in Fig. 1, a sample is pumped without prior pH adjustment through a column array consisting of XAD-8, MSC-1 hydrogen-saturated cation-exchange resin, and A-7 resin in the free-base form. The majority of the organic acids in the sample, such as fulvic acid, sorb onto the A-7 resin. These acids are eluted from the column by adjusting the influent pH to 11.5. The cation-exchange resin, MSC-1, is essential for the collection of organic acids on the A-7 resin. Sorption of the acids on the A-7 resin is by hydrogen bonding and ion exchange. Both of these mechanisms are pH dependent, and efficient sorption is favored at acidic pHs. The cation-exchange resin serves two functions: (a) all of the organic acids are in the hydrogen form passing through the cation-exchange resin, and (b) the pH of the effluent from the cation-exchange resin is low. XAD-8 absorbs the hydrophobic neutral compounds and is not a necessary part of the system if organic acids are the only solutes of interest.

AREAS OF CONCERN

The development of chromatographic methods has greatly increased the efficiency of the isolation process for aquatic humic substances and has been responsible, in part, for the increased interest over the past two decades in the chemistry of these materials. However, proper application of the methods requires an understanding of the underlying principles and an appreciation of the limitations of each method. Some of the key areas of concern related

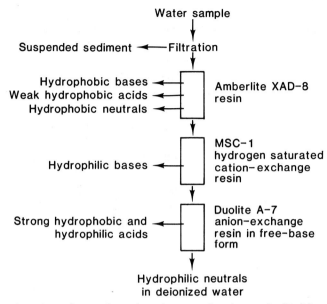

Fig. 1—Fractionation of organic carbon in water by the method of Leenheer and Noyes (1984).

to the isolation of humic substances from water will be addressed in this section.

Operational definition

One of the difficulties encountered in the study of humic substances relates to terminology. Humic substances do not conform to a unique chemical entity and cannot be described in unambiguous structural terms. As a result, humic substances are described operationally, representing a complex mixture of organic acids. The terms used to describe humic substances, such as humic and fulvic acid, originate in the field of soil science where they are used to describe distinct fractions of soil organic matter that are extracted from soil by rather severe chemical treatment. As pointed out by Reuter and Perdue (1981), standard soil extraction techniques are unlikely to yield dissolved humic substances representative of the fraction of soil humus mobilized by meteoric waters.

It would be more reasonable to classify these aquatic materials based on the chromatographic methods used to extract them from water. However, problems exist with this method of terminology as well. For instance, according to Malcolm (1985), aquatic humic substances are defined as comprising that portion of the nonspecific, amorphous, carbonaceous fraction of the dissolved organic carbon in water that, upon acidification to pH 2, has a column distribution coefficient of greater than 50 on XAD-8 resin at 50 percent breakthrough of the column. In terms of the fractionation scheme of Leenheer (1981), this fraction of material represents the hydrophobic acid fraction of the dissolved organic carbon pool. The most serious problem with these definitions is that no consensus exists among those working in the field as to the method of extraction; therefore, no consensus exists as to what part of the dissolved organic carbon pool is classified under the definitions of fulvic or humic acids.

Presently, a number of variations of resin isolation procedures are employed to isolate aquatic humic substances. The fractionation of the dissolved organic carbon in a sample varies depending on the type of resin used, the amount of sample passed through the resin, and the eluents that are used. Consequently, samples collected using different methods are not directly comparable. In addition, the capacity factor cutoffs for organic solutes on XAD resins are also dependent on the concentration of material in the sample. The fractionation of organic solutes on XAD resins has been defined for samples in the concentration range of 2–10 mg C/L, but is not well understood for samples with higher concentrations of organic carbon (>35 mg C/L). Therefore, it is difficult to obtain consistency in comparing materials extracted from sample sites that vary greatly in the concentration of organic matter. The problem of variability in the procedures employed by different researchers for isolating aquatic humic substances and the uncertainties in the procedures for systems of high DOC have implications for establishing structural models and defining the chemistry of these materials. For example, the aromatic content of aquatic fulvic acids depends in part on the fractionation of the dissolved organic carbon, with hydrophilic acids having a lower content than hydrophobic acids (Fig. 2). The debate concerning the "true aromaticity" of aquatic humic substances may be due, in part, to variations in isolation procedures.

Other problems arise from the use of chromatographic fractionation to define these materials. Unfortunately, the separation of different fractions on resin sorbents is not sharp, resulting in overlap between fractions. For instance, a certain amount of the hydrophilic acids in the sample will be concentrated along with the hydrophobic acid fraction. These compounds are eluted in the same manner as humic substances and will be included in the final extract.

Similarly, compounds that comprise the hydrophobic neutral fraction can be included in the hydrophobic acid fraction during the elution of the acids. The hydrophobic neutral fraction includes compounds such as long-chain fatty acids with fewer than one carboxylic acid functional group per 10 carbon atoms, hydrophobic surfactants such as alkylbenzene sulphonates and linear alkylbenzene sulphonates in sewage effluents, and long-chain alkylbenzenes

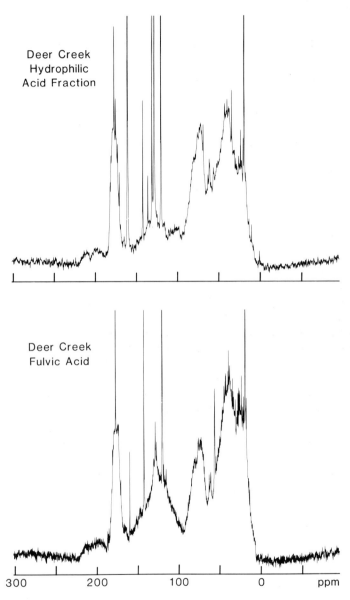

Fig. 2—^{13}C NMR spectra of fulvic acid and the hydrophilic acid fraction isolated from Deer Creek, Colorado. NMR conditions were as follows: 30,000 Hz spectral width, 45° pulse angle, 0.2 s acquisition time, 1.0 s pulse delay, continuous decoupling, 10 Hz line broadening.

from the production of these surfactants (Eganhouse 1986). While many of these compounds are not neutrals in the classical sense, they are operationally included in the hydrophobic neutral fraction becuase of the nature of their interations with the resin. The sorption of these compounds on resins such as the XAD resins is not pH dependent. Hydrophobic neutrals absorb on the resin as the sample is processed and should be separated from the humic substances during the elution step if elution of the acids is brought about by pH adjustment. When the concentration of hydrophobic acids being eluted from the resin is high, however, the acids can interact with the neutral compounds to help solubilize them and incorporate them into the final extract. Consequently, it is difficult to obtain a "pure" hydrophobic acid fraction and there may be some hydrophobic neutrals included in the isolate. Harvey (1985) has reported the presence of such compounds in marine humic substances isolated on XAD-2.

Compounds that comprise the hydrophobic neutral fraction are eluted easily with organic solvents, such as methanol or acetonitrile. If the hydrophobic acid fraction is eluted with organic solvents, these compounds will be included in the extract to a much greater degree than if pH adjustment is used to elute the hydrophobic acids. The resulting sample may contain lipid-like compounds or alkylbenzenes that appear to be structural components of the humic substances. Therefore, if an organic solvent is to be utilized to elute humic substances, a two column system needs to be used to separate these two fractions. A precolumn of XAD resin can be used to remove the hydrophobic neutral fraction from the sample before the pH of the sample is adjusted for sorption of the organic acids on either another column of XAD resin or on A-7. In the method of Leenheer and Noyes (1984), XAD-8 is used to retain the hydrophobic neutral fraction in this manner (Fig. 1).

Another complication arising from an operational definition based on the chromatographic fractionation of dissolved organic carbon is that a large amount (up to 20% in some samples) of humic-like compounds is fractionated into the hydrophilic acid fraction. Consequently, this fraction of the DOC pool is rarely isolated and is, therefore, often ignored in the study of aquatic humic substances. However, the similarity of this fraction to aquatic fulvic acid is evident in the ^{13}C NMR spectra for an aquatic fulvic acid and the corresponding hydrophilic acid from Deer Creek, Colorado given in Fig. 2. These samples were obtained by first isolating the fulvic acid on XAD-8 and then isolating the hydrophilic acids from the XAD-8 effluent on XAD-4. The spectra are nonquantitative because differential saturation effects and differential nuclear Overhauser effects have not been eliminated; however, the spectra for both samples were obtained under the same acquisition parameters. Although the fulvic acid sample was found to be more aromatic than the hydrophilic acid sample (110–160 ppm), the spectra indicate that these samples are similar with respect to other regions in the spectra. Of particular environmental significance is the amount of carboxylic acid functional group content in the hydrophilic acid fraction (160–190 ppm). These compounds may be of geochemical significance in the environment, however, they have been largely ignored in the study of aquatic humic substances.

Sample alteration

A significant area of concern in isolating humic substances from water is the potential for chemical alterations occurring in a sample as a result of the isolation procedures themselves. The commonly used chromatographic methods for isolating these materials require that the sample come in contact with acid, base, or organic solvents. These chemicals can interact with humic substances to alter permanently the chemical structure of the isolates.

Ester hydrolysis is of particular concern in the isolation of humic substances. The hydrolysis of organic esters is pH dependent and can occur in both acidic and basic solutions. In general, the most significant acid–catalyzed hydrolysis of organic esters occurs at very high concentrations of acid (<pH 1) (Kirby 1972). In the case of acetylsalicylic acid, for instance, the lowest overall rate of hydrolysis is at pH 2, with the highest rates of acid-catalyzed hydrolysis occurring at pH 0 (Garrett 1957). As part of the extraction procedure for aquatic substances, pH values of 1–2 are often encountered. At these pH values the amount of acid-catalyzed hydrolysis for organic esters should be negligible, with the rates being less than or comparable to rates of hydrolysis at pH 4–8. There are, however, certain organic-inorganic esters, such as sulfate esters, that are very labile at low pH values (Bierderbeck 1978), and it is almost certain that these esters, if present, would be irreversibly hydrolyzed during the isolation of aquatic humic substances.

The effects of base hydrolysis on humic substances are potentially much more serious. Ester hydrolysis in the presence of base has been reported for soil humic substances by Schnitzer and Neyroud (1975) and Gregor and Powell (1987), and for aquatic humic substances by Liao et al. (1982) and Bowles et al. (in press). During the isolation of aquatic humic substances on either the XAD resins or Duolite A-7, the absorbed material is often eluted with 0.1 N NaOH resulting in a solution pH of 13. At this pH ester hydrolysis is very rapid, and despite efforts to minimize hydrolysis at this step labile ester groups will react, resulting in irreversible hydrolytic reactions. In the event of ester hydrolysis, the isolated material is not truly representative of the original fulvic acid. Gregor and Powell (1987) report that alkali treatment of soil fulvic acids, extracted with acid phosphate, resulted in a net generation of titratable carboxyl groups from the hydrolysis of esters and an increase in the relative binding strength for copper (II) ions consistent with the increase of potential binding sites arising from ester hydrolysis. This presents serious implications in interpreting the geochemical and environmental behavior of aquatic humic substances, such as the ability to bind metal ions. The problem of organic ester hydrolysis potentially can be avoided by eluting with a solvent such as CH_3CN, or by using another method that avoids the use of base.

A similar set of complications is encountered when CH_3OH is used to elute humic substances from resin sorbents such as XAD resins. Aside from the problem of incorporating solutes comprising the hydrophobic neutral fraction into the isolated humic substances, methanolysis reactions can occur with both carboxylic acid functional groups and with organic esters. The esterification of carboxylic acid functional groups is catalyzed by acids, but not by bases; therefore, the problem can be avoided. However, alcoholysis of esters is both acid and base catalyzed and results in a transesterification reaction. As pointed out by Steelink (1985), an example of the type of reaction that can occur was published by Armitage et al. (1961) for the methanolysis of Chinese tannin, in which ester bonds were hydrolyzed by methanol resulting in the formation of the corresponding methyl esters. When treated with methanol, the original Chinese tannin yielded gallic acid derivatives plus a smaller tannin molecule. In the case of humic substances, there is the potential for transesterification in the presence of alcohol and subsequent sample degradation. Therefore, contact with alcohol needs to be avoided.

Ammonium hydroxide has been used as an eluent for aquatic humic substances from XAD resins (Mantoura and Riley 1975) and also interacts strongly with humic substances. Interactions of ammonia with soil organic matter have been studied for many years (Stevenson 1982). In addition to aminolysis reactions with organic esters in the presence of base, NH_3 can interact with other components in humic substances such as phenols, and is difficult to eliminate completely from the final isolate. Stepanov (1969) found that nonexchangeable absorption of nitrogen and its fixation by soil humic acid followed the reaction of humic acid with NH_4OH. Evidence also exists for the incorporation of nitrogen into aquatic humic substances. Harvey and Boran (1985) report that marine fulvic acids isolated with NH_4OH had nitrogen contents as high as 3–5%. These values are higher than found for other aquatic fulvic acids isolated without NH_4OH. While the nature of this nitrogen is unknown, use of NH_4OH as an eluent for aquatic humic substances will result in an erroneously high nitrogen content in the isolate.

Up to the present time (1986), almost all resin-based extraction schemes have involved contact of the sample with one or more of the reagents discussed previously. The samples extracted in this way are unlikely to have survived the process unaltered, and the extent to which these samples are representative of the original organic matter is uncertain. In isolating these materials, some degree of alteration is unavoidable; however, this alteration needs to be minimized and acknowledged if subsequent analyses are to shed light on the biogenesis and environmental roles of these materials.

Sample contamination

The incorporation of impurities into isolated aquatic humic substances is another potential problem when these materials are isolated with resin sorbents, such as XAD resins or Duolite A-7. These synthetic materials, as delivered from the manufacturer, contain residual monomers, artifacts of the polymerization pathway, and chemical preservatives. Extensive cleanup procedures involving Soxhlet extraction are required to eliminate these components from the resin. However, to eliminate all the bleed is impossible when using resin sorbents, and to identify the bleed components from whatever resin is used is essential.

In general, the bleed characteristics of the styrene-divinylbenzene XAD resins, such as XAD-2, are much better than the acrylic-ester resins such as XAD-8. Despite exhaustive cleaning of the resin before use, XAD-8 bleeds on the order of 3 mg C/L in 0.1 N NaOH. A major component of the bleed is methyl acrylic acid, which has a low capacity factory on XAD-8 at pH 2 and can be separated from the sample to a large extent during successive reconcentration steps. Similar bleed contamination occurs with the anion-exchange resin Duolite A-7. This resin is a phenol-formaldehyde resin, and bleed concentrations on the order of 30 mg C/L have been observed in the elution of A-7 at pH 11.5. This bleed is predominantly amino phenols, formic acid, and formaldehyde. Formic acid and formaldehyde are volatile and are removable by evaporation, and the nonvolatile amino phenols can be removed by a hydrogen-saturated cation-exchange resin.

All resin bleed components cannot be eliminated from the isolated materials. Some sharp lines, present in the NMR spectra for the fulvic acid and hydrophilic acids presented in Fig. 2, can be accounted for by the presence of resin bleed in the samples. Although the resins used to obtain these samples had been extensively extracted, both samples contain methyl acrylic acid from the XAD-8 resin. This problem is more significant when the concentration of humic substances in the original water sample is low, since the amount of bleed relative to the amount of isolated material will be greater.

SUMMARY AND CONCLUSIONS

The ultimate goal in the study of aquatic humic substances is to relate the structural and chemical information obtained on humic substances to their biogenesis and environmental roles. To accomplish this end, aquatic humic substances must first be isolated from the environment and subjected to chemical studies to define adequately and accurately their structures and chemical properties. In order that the conclusions drawn from these studies truly reflect the processes in the natural system, it is of great importance that the compounds under study are representative of those in the environment.

In the environment these compounds are part of a complex chemical system. Unfortunately, the isolation process itself brings about unavoidable alterations in the chemistry of these materials. Included in these alterations are the disruption of interactions of humic substances with metal ions, dissolved silica, clay colloids, and other organic species. In many instances these alterations are reversible and do not present problems in subsequent analyses. Other alterations, such as ester hydrolysis or methanolysis, result in nonreversible chemical changes that alter the chemical properties of the isolated material and may lead to incorrect conclusions. The potential also exists that other chemical constituents will be inadvertently included in humic substances either in the form of other fractions of organic matter that are not adequately separated from the dissolved humic substances or as resin bleed. Each of these problems can be a hindrance in understanding the roles played by humic substances in the environment; however, each of these problems can be minimized by careful analysis of the isolation procedure.

The chromatographic methods utilizing Duolite A-7 or the XAD resins are the most efficient methods for isolating humic substances from water. These methods make it possible to process large volumes of water and to obtain a sufficient amount of material to characterize adequately the humic substances. It is important to recognize that some alteration of the isolated humic substances is unavoidable. However, isolation procedures can be designed that minimize the degree of alteration and contamination by understanding the basic principles of the isolation process.

Acknowledgements. The author is indebted to K. Thorn for providing ^{13}C NMR spectra presented in this manuscript, and to R. Antweiler, K. Thorn, and R. Wershaw, all of the U.S. Geological Survey, for their critical and helpful review comments.

REFERENCES

Abrams, I.M. 1975. Macroporous condensate resins as adsorbents. *Indus. Eng. Chem. Prod. Res. Dev.* **14**: 108–112.

Aiken, G.R. 1985. Isolation and concentration techniques for aquatic humic substances. In: Humic Substances in Soil, Sediment, and Water, eds. G.R. Aiken, D.M. McKnight, R.L. Wershaw, and P. MacCarthy, pp. 363–386. New York: John Wiley and Sons.

Aiken, G.R. 1987. Isolation of organic acids from large volumes of water by adsorption on macroporous resins. In: Organic Pollutants in Water, vol. 214, Advances in Chemistry Series, eds. I.H. Suffet and M. Malaiyandi, pp. 295–308. Washington,

D.C.: American Chemical Society.

Aiken, G.R.; Thurman, E.M.; Malcolm, R.L.; and Walton, H.F. 1979. Comparison of XAD macroporous resins for the concentration of fulvic acid from aqueous solution. *Analyt. Chem.* **51**: 1799–1803.

Armitage, R.G.S.; Gramshaw, J.W.; Haslam, E.; Haworth, R.D.; Jones, K.; Rogers, H.J.; and Searle, T. 1961. Gallotannins, part III. The constitution of Chinese, Turkish, Sumach, and Tara Tannins. *J. Chem. Soc.* **1842**:

Bierderbeck, V.O. 1978. Soil organic sulfur and fertility. In: Soil Organic Matter, eds. M. Schnitzer and S.U. Khan, pp. 273–310. New York: Elsevier Publishing Company.

Bowles, E.C.; Antweiler, R.C.; and MacCarthy, P. 1988. Acid-base titration and hydrolysis of Suwannee River fulvic acid. U.S. Geological Survey Water Supply Paper. Washington, D.C.: U.S. Govt. Printing Office, in press.

Cheng, K.L. 1977. Separation of humic acid on XAD resins. *Microchim. Acta* **1977–II**: 389–396.

Danielsson, L.G. 1982. On the use of filters for distinguishing between dissolved and particulate fractions in natural waters. *Water Res.* **16**: 179–182.

Eganhouse, R.P. 1986. Long-chain alkylbenzenes: their analytical chemistry, environmental occurrence and fate. *Int. J. Env. Analyt. Chem.* **26**: 241–263.

Garrett, E.R. 1957. The kinetics of solvolysis of acyl esters of salicylic acid. *J. Am. Chem. Soc.* **79**: 3401–3408.

Gregor, J.E., and Powell, H.K.J. 1987. Effects of extraction procedures on fulvic acid properties. *Sci. Total Env.* **62**: 3–12.

Harvey, G.R. 1985. On the origin of alkylbenzenes in geochemical samples. *Marine chem.* **16**: 187–188.

Harvey, G.R., and Boran, D.A. 1985. Geochemistry of humic substances in seawater. In: Humic Substances in Soil, Sediment, and Water, eds. G.R. Aiken, D.M. McKnight, R.L. Wershaw, and P. MacCarthy, pp. 181–210. New York: John Wiley and Sons.

Kirby, A.J. 1972. Hydrolysis and formation of esters of organic acids. In: Comprehensive Chemical Kinetics, eds. C.H. Bamford and C.F.H. Tipper, vol. 10, pp. 57–208. London: Elsevier Publishing Company.

Leenheer, J.A. 1981. Comprehensive approach to preparative isolation and fractionation of dissolved organic carbon from natural waters and wastewaters. *Env. Sci. Tech.* **15**: 578–587.

Leenheer, J.A., and Noyes, T.I. 1984. A filtration and column adsorbent system for onsite concentration and fractionation of organic substances from large volumes of water. U.S. Geological Survey Water Supply Paper Number 2230. Washington, D.C.: U.S. Government Printing Office.

Liao, W.; Christman, R.F.; Johnson, J.D.; and Millington, D.S. 1982. Structural characterization of aquatic humic material. *Env. Sci. Tech.* **16**: 403–410.

Mantoura, R.F.C., and Riley, J.P. 1975. The analytical concentration of humic substances from natural waters. *Analyt. Chim. Acta* **76**: 97–106.

Malcolm, R.L. 1985. Geochemistry of stream fulvic and humic substances. In: Humic Substances in Soil, Sediment, and Water, eds. G.R. Aiken, D.M. McKnight, R.L. Wershaw, and P. MacCarthy, pp. 181–210. New York: John Wiley and Sons.

Packham, R.F. 1964. Studies of organic colour in natural water. *Proc. Soc. Water Treat. Exam.* **13**: 316–334.

Reuter, J.H., and Perdue, E.M., 1981. Calculation of molecular weights of humic substances from colligative data: application to aquatic humus and its molecular weight fractions. *Geochim. Cosmochim. Acta* **41**: 325–334.

Schnitzer, M., and Neyroud, J.A. 1975. Alkanes and fatty acids in humic substances. *Fuel* **54**: 17–19.

Steelink, C. 1985. Implications of elemental characteristics of humic substances. In: Humic Substances in Soil, Sediment, and Water, eds. G.R. Aiken, D.M. McKnight, R.L. Wershaw, and P. MacCarthy, pp. 457–476. New York: John Wiley and Sons.

Stepanov, V.V. 1969. Reaction of humic acids with some nitrogen containing compounds. *Pochvovedeniye* **3**: 37–43.

Stevenson, F.J. 1982. Humus Chemistry. New York: Wiley-Interscience.

Thurman, E.M.; Malcolm, R.L. 1981 Preparative isolation of aquatic humic substances. *Env. Sci. Tech.* **15**: 463–466.

Thurman, E.M.; Malcolm, R.L.; and Aiken, G.R. 1978. Prediction of capacity factors for aqueous organic solutes on a porous acrylic resin. *Analyt. Chem.* **50**: 775–779.

Standing, left to right:
George Aiken, Fauzi Mantoura, Alfons Hack, Walter Fischer,
Mike Thurman

Seated (center), left to right:
Frank Stevenson, Roger Swift, Barbara Szpakowska, John Parsons

Seated (front), left to right:
Marc Ewald, Roger Pocklington, Uli Förstner

Humic Substances and Their Role in the Environment
eds. F.H. Frimmel and R.F. Christman, pp. 31–43
John Wiley & Sons Limited
© S. Bernhard, Dahlem Konferenzen, 1988.

Isolation of Soil and Aquatic Humic Substances
Group Report

E.M. Thurman, Rapporteur
G.R. Aiken
M. Ewald
W.R. Fischer
U. Förstner
A.H. Hack
R.F.C. Mantoura (Moderator)

J.W. Parsons
R. Pocklington
F.J. Stevenson
R.S. Swift
B. Szpakowska

INTRODUCTION

When considering the isolation of humic substances, one is immediately confronted with the question, "What are humic substances?" Although this is not an original question, our ideas on isolation and fractionation are preconditioned by this question. In fact, the following section of this chapter deals with the terminology of humic substances and a philosophical discussion of the importance of terminology.

The second question our group addressed was, "Are extractions necessary?" Because humic substances are operationally defined by extractions, the second question is closely related to the first.

Although we all think extractions are necessary, there is a growing interest in direct analysis of soil and water that bypasses the need for humic extraction. In fact, there is a minority view that specific compound or specific class analysis (pigments, lipids, etc.) is the only way to determine the structure of humic and fulvic acids. Furthermore, this chapter examines soil and aquatic isolation procedures and tries to establish simple guidelines for the terms humic and fulvic acid. There is discussion of the problems with our methods of isolation. Next, the chapter considers several possible methods of isolation and fractionation on *soil* and *aquatic* humic substances. Finally, our chief accomplishments in this chapter are two. First, the agreement among soil, freshwater, and marine chemists to a mutually acceptable nomenclature for humic substances is of the utmost importance. Second, the

31

relationship of marine "Gelbstoffe" to yellow organic acids of freshwaters, and of this to the fulvic/humic fractions of soil is under investigation; thus, until we know that they are identical we must use "environmental descriptors" to distinguish the source of our fractions. This effort is directed at the terms, not for their own value, but to clarify present and future results among soil, marine, and freshwater chemists who work on humic substances.

HUMIC ISOLATION: TERMINOLOGY

The terms

The terminology of humic chemistry has a long history dating from the early work of Achard (1786) who extracted peat with alkali to yield a dark precipitate that we would now call "humic substances." Although humic terminology has evolved through an interesting course over 200 years, soil scientists generally agree on four terms:

1) humic substances: a general category of naturally occurring, biogenic, heterogeneous organic substances from soil that can be characterized as being yellow to black in color, of high molecular weight, and refractory (glossary, Aiken et al. 1985);
2) humic acid: a fraction of humic substances that is soluble at alkaline pH but precipitates at acidic pH;
3) fulvic acid: a fraction of humic substances that is water soluble at alkaline and acidic pH;
4) humin: a fraction of humic substances that is insoluble at any pH.

With this general terminology for soil science, the study of humic substances has proceeded to characterize soils from many environments and much is known of the nature of soil humus (Schnitzer and Khan 1972; Stevenson 1982; Aiken et al. 1985). Approximately 180 years ago Berzelius (1806) isolated yellow-colored organic acids from a spring near Porla, Sweden, noting that these acids were similar to those from soil and probably were washed from soil as their salts. Berzelius called these acids crenic and aprocrenic acids, which mean organic acids from springs. Waksman (1938) studied the formation of humus in water and distinguished among humic substances from seawater, rivers, and lakes on the basis of origin and chemical analyses. Original work on the nature of organic matter in seawater includes Brandt (1899), Kalle (1937, 1966), Fogg (1966), and Nissenbaum and Kaplan (1972). Kalle (1937, 1966) isolated yellow organic acids from seawater that he named "Gelbstoff." Shapiro (1957) coined the term "yellow organic acids" for the dissolved organic matter in a pond in Connecticut, U.S.A. Over the past 20 years, with the interest in both the chemistry of seawater and freshwater, there have been numerous studies of the nature of these natural yellow-colored organic acids (Gjessing 1976; Aiken et al. 1979; Harvey et al. 1983; Wilson et al. 1983; Ertel and Hedges 1984) and over this time period there has been a borrowing of the terms humic and fulvic acid from the soil literature that has led to some confusion, which is far from trivial (Pocklington 1977).

Because of the environmental effects of aquatic organic acids, including trace metal complexation, trihalomethane production, and pesticide partitioning, many researchers have used *soil* humic substances as analogs of aquatic yellow organic acids, which could lead to misinterpretation of scientific literature (Malcolm and MacCarthy 1986). Bearing these ideas in mind we set out to establish a consistent set of terms for humic substances from various environmental matrices that would allow scientists in other disciplines to use soil terminology with a meaning that is consistent with the soil literature. When natural substances are not fully defined in a chemical sense, consistent use of terminology is initially important.

The concept

Figure 1 is not intended as an isolation scheme, rather it describes our conceptual view of humic and fulvic acids. We strongly emphasize that this figure be used for the concepts only. The right side of the figure is for solid materials (soil, sediments, etc.) and the left side is for aquatic samples (freshwater, marine, pore waters, etc.). During steps I and II (Pretreatment and Extraction) samples are treated in different ways. In step III (Acidification) the procedures join and move toward a common terminology. It is the joining at step III that permits the use of the terms humic and fulvic acid for samples from many different environments.

We consider this combination of soil, marine, and freshwater terminology for humic substances and the use of environmental descriptors or source indicators (e.g., *soil, aquatic, sedimentary* humic substances) the major accomplishment of our sessions at the Dahlem Workshop.

Another important concept is illustrated in step III, which describes the generation of the humic and fulvic acid fractions. These two fractions are not solely humic and fulvic acids but may contain specific compound classes such as polysaccharides and amino acids, which are removed in the purification step. A source indicator is used to clarify the origin of the sample (e.g., *soil* fulvic acid fraction, *aquatic* humic acid fraction).

Step IV describes the "purification" of the humic and fulvic acid fractions to yield the products, humic and fulvic acids, and step V describes a verification step. Verification may be viewed as the research that we do to identify the product of our isolation procedure. Verification may be simple, such as elemental analysis and acidity, or it may be more complex such as ^{13}C NMR, molecular weight determination, etc.

A Guide to Nomenclature of Humic Substances

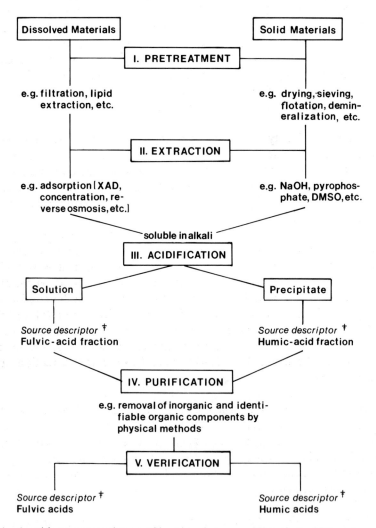

Fig. 1—A guide to nomenclature of humic substances. Note that additional analytical guidelines for verification would include comparisons with average elemental, ultraviolet-visible absorbance, and acidity properties of various humic fractions obtained from critical compilation of the literature.

The source from which the sample was isolated (e.g., *soil*, *aquatic*, and *sedimentary*, etc.). Thus, fulvic acid isolated from seawater would be called "*marine aquatic* fulvic acid."

The isolation of a sample

Pretreatment.   This step involves preparation of the sample for the extraction of humic and fulvic acids. For soil substances, typical pretreatments include density separation of plant fragments, drying, sieving, and demineralization by HF and HCl extraction. Water samples, on the other hand, are commonly filtered and acidified with HCl to pH 2.0. If contaminants are present in the sample, they may be solvent or resin extacted at neutral pH. Samples from other sources such as sediments may receive different pretreatment procedures, e.g., freeze-drying.

Extraction.   The typical extraction for soil involves the removal of an alkaline soluble fraction. Commonly used solvents include NaOH and pyrophosphate. Organic solvents such as dimethyl sulfoxide and methylisobutylketone, may also be used. Although these organic-solvent extractions are not in themselves alkaline, they yield an organic matter fraction that is soluble in alkali solutions.

Water samples can be treated by column sorption techniques (e.g., XAD-2 or XAD-8 resins) or by other concentration mechanisms such as ion exchange, freeze-drying, roto-evaporation, etc. Although it is possible that many other procedures may be tested in this isolation scheme, at the moment we consider that the isolation scheme of the International Humic Substances Society (IHSS) is useful for soil, marine, and freshwater samples (Aiken et al. 1985, pp. 8–9). For marine samples, XAD-2 resin is considered to be equivalent to XAD-8 in the isolation scheme. Additionally, soil fulvic acid solutions may be isolated on XAD resin. Criteria for extraction are also given by Parsons (this volume).

Acidification.   The alkali-soluble extract is acidified to pH ≤ 2.0 with HCl and allowed to precipitate. Chilling the samples hastens the process. Centrifugation is needed to separate the sample into two fractions, a fulvic acid and a humic acid fraction. For those interested in the details of pretreatment, extraction, and acidification, see Hayes (1985) and Aiken (1985) for soil and aquatic samples.

Purification.   Since the structures of humic substances are not known precisely, the term "purification," as used here, does not have its usual meaning. Rather our usage refers to the removal of known nonhumic inorganic and organic materials by physical methods. Desalting is a common procedure for both *aquatic* and *soil* humic and fulvic acids and detailed procedures may be found in Hayes (1985) and Aiken (1985).

Purification is needed to remove substances that are physically but not covalently bound in humic extracts (salts, metals, lipids, polysaccharides, fatty acids, etc.).

Verification.   The final step in the isolation of humic and fulvic acids is the verification of the isolate as humic and fulvic acid. This is done by comparing the composition of the isolate with average elemental and acidity properties (see footnote in Fig. 1). In fact, the extensive characterization that may be done is commonly the focus of research on humic structures. We note with pleasure that the IHSS plans to publish values and ranges for average soil and aquatic humic and fulvic acids that are part of its collection (Aiken et al. 1985, pp. 8–9).

EXPERIMENTAL CONSIDERATIONS

Soil

During the isolation of humic substances from soil, there are a number of alterations that may occur. They include base hydrolysis of esters and incorporation of oxygen into the humic structure under alkaline conditions. It has also been noted that ammonia may be released during extraction of humic material by alkali (Parsons, this volume). Wilson has noted the appearance of formic and other organic acids by ^{13}C NMR over a 5 day period, when humic substances are allowed to stand in alkaline solution (Wilson 1987). These effects may be mitigated by minimizing the time of contact between the humic isolate and the aqueous base and by extracting under nitrogen gas (Hayes 1985). Samples should never be stored at alkaline pH, and the practice of storing samples in amber glass bottles should be followed.

Solutions containing ammonia or reactive nitrogen should not be used as a humic solvent because of the possibility of crosslinking the nitrogen to phenol and quinone structures. Much greater nitrogen contents (factor of two) have been reported for samples extracted with ammonia. Freeze-drying may also alter a humic structure by forming internal esters. Hydrogen from a hydroxyl group and a hydroxyl from a carboxyl group (H-OH) are removed, which decreases subsequent water solubility and may alter structure by crosslinking the molecule. This type of change may be reversed slowly in base. Finally, if freeze-drying is done in an excess of acid, charring of the sample may occur. As a result of these considerations, several recommendations for isolation of humic substances from soil include:

1) minimizing damage to humic molecules in any extraction, especially oxygen incorporation during alkaline extraction;
2) using short extraction times in base (10–20 minutes);
3) continuing research on comparison of soil extraction techniques with organic solvents (dipolar aprotic) and the artifacts created by extraction;
4) milder, serial extractions are recommended;
5) exploring the use of supercritical fluids for extraction

(such as carbon dioxide);

6) storing humic substances in amber bottles away from light.

Water

When isolating humic substances from water there are several precautions needed to avoid alteration. Because alkaline extraction is typically used to remove humic substances sorbed to XAD or ion exchange resins, contact time with hydroxide ion should be minimized. One way to accomplish this is to elute the humic substances from the XAD resin and desalt directly onto a cation-exchange resin in the hydrogen-saturated form. Another possibility is to elute the humic substances with sodium bicarbonate at pH 8.0, or tetraethylamine, acetonitrile, etc. Phenolic hydroxyls are usually not ionized at this pH, but because *aquatic* humic substances contain a much lower phenolic content than *soil* humic substances (Thurman 1985), this is not an important consideration. The sodium bicarbonate may be rapidly removed by cation exchange and degassed as carbon dioxide.

Acid may also hydrolyze labile ester linkages that occur between sugars and humic substances, as occurs in certain natural products. Because it is difficult to isolate humic substances from most natural waters without pH adjustment, it is difficult to assess the damage done to the humic molecules.

Few studies have addressed alteration products of an aquatic isolation procedure. When possible it is desirable to make *in situ* measurements of humic properties that can later be compared with results of the freeze-dried product. An example would be the content of labile sugars in the humic molecule, which could be analyzed in the column effluent, in the original sample, and in the humic extract.

Important contaminants created during the isolation of *aquatic* humic substances are degradation products of the resins or solvents used to isolate the humic material. XAD-8 is known to bleed or hydrolyze as a result of the ester structure of the resin. Care in cleaning the resin immediately before use is necessary (Thurman and Malcolm 1981) and is especially important for those involved in environmental toxicology. Fortunately, many of the products of resin bleed, such as methyl acrylic acid, are removed in freeze-drying. More research is needed both in purifying the resins and in searching for resins that are more chemically stable.

Recommendations for isolation of humic substances from water include:

1) the use of resin isolation methods that do not require either acid or base (weak ion exchange);
2) the investigation of other types of isolation (ultrafiltration, continuous ultracentrifugation, porous glass filtration, see next section);
3) minimizing contact of sample with either acid or base during isolation

stage by use of cation exchange or acidification within minutes of extraction;

4) minimizing resin bleed with careful cleanup procedures or removal of products from extracts;

5) investigating the use of inorganic ion exchange resins that are selective for aquatic humic substances;

6) studying elution properties of other solvents such as acetonitrile, tetraethylamine, etc.

FUTURE RESEARCH IN ISOLATION AND FRACTIONATION

Soil

Perhaps the most exciting new approach in the examination of soil humic substances is to study the soils intact without extraction. Possible new methods include solid state ^{13}C NMR on the whole soil. This may be preceded by density and size fractionation. Work by Wilson (1987) has shown that the structure of organic matter in different size fractions from the same soil varies considerably. The organic matter in the largest size fraction (20–250 μm) contained alkyl, carbohydrate, and carboxylic carbon. The middle size fractions were highly aromatic with many carboxylic groups. The organic matter in the smallest size fraction (≤ 0.2 μm) was highly aliphatic.

Other techniques that are useful for solid analysis are Fourier transform infared spectrometer (FT-IR) microscopy and new NMR techniques such as dynamic nuclear polarization from electrons. This latter technique may improve sensitivity from 3% to 0.01% carbon. ^{15}N and Si NMR are also possible new techniques (Wilson 1987).

Supercritical fluid (SF) extraction of soils may be an emerging technique for the extraction of *soil* humic substances, especially for the insoluble fractions such as humin. Generally, SF extraction is applicable to nonionic compounds, such as polycyclic aromatic compounds of high molecular weight (>300) and heat labile compounds. Ionic compounds are not soluble in supercritical fluids. It may be possible to hydrogen saturate the soil in order to improve the extraction of *soil* humic substances by SF. Probably one of the most exciting advances in SF chromatography (SFC) has not yet been marketed. Lee Scientific (Salt Lake City, Utah, U.S.A.) is working in conjunction with Finningan (Palo Alto, California, U.S.A.) to interface the SFC with a quadrupole mass spectrometer (SFC-MS). Furthermore, one could easily imagine combining the SF extractor with the SFC-MS to have the entire system in one unit.

Thus, the soil sample could be placed into the extractor and the pressure gradient programmed to elute individual classes from soil, while at the same time separating them in the high performance capillary chromatograph and obtaining mass spectra. Although such a procedure is not yet complete, it is likely that the SFC-MS will be introduced at the Pittsburgh Conference in New Orleans (U.S.A.) in March of 1988. Furthermore, nondestructive detectors for gas chromatography, such as FT-IR, could be put ahead of the mass spectrometer.

There are a number of organic solvents that are currently evaluated for the extraction of humic substances without alteration; for example, dipolar, aprotic solvents such as dimethyl sulfoxide, methylisobutylketone, and diethyl carbonate may be used and combined with normal phase liquid chromatography.

Advances in liquid chromatography include high resolution microbore liquid chromatography with both normal and reverse phases. Crown ethers and ion pair reagents (tetrabutyl ammonium phosphate) may be used to solubilize ionic humic substances and permit the use of normal phase chromatography. This same technique would work well with soil humic substances, especially given the large quantities of *soil* humic substances that are available for extraction.

There have been numerous advances in ion chromatography and in packing materials that use anion exchange as a separation technique. This might be used to separate the most soluble of the *soil* fulvic acid fractions. Dionex has developed a new high performance liquid chromatograph that uses all plastic columns, 5 μm styrene-divinylbenzene packings, and teflon tubing that will operate at elevated pressure (140–280 kg/cm^2) and is completely automated. Because the system contains no metal, both acid and alkali can be effectively used for gradient elutions. This should provide a powerful technique for fractionation of *soil* and *aquatic* organic matter.

Specific extraction of soils is yet another way to achieve isolation and fractionation at the same time. For example, pyrophosphate has been used with 8 mol/l urea to remove metal-humate complexes.

Finally, new separation techniques that might be used on soil and water include a) field flow fractionation (Dupont, U.S.A.), b) continuous flow ultracentrifugation (Sorval, U.S.A.), tangential flow filtration (Millipore, U.S.A.), and c) spray drying, electrodialysis, and porous capillary glass ultrafiltration (Schott, Mainz, F.R.G.). These methods have been used on biological and food products and should be applicable to colloidal solutions of *soil* humic substances.

Water

Major advances in the isolation of *aquatic* humic substances occurred in the mid-1970s when XAD resins were developed and used to study marine organic matter (Riley and Taylor 1969; Mantoura and Riley 1975). The resins have been developed and tested extensively since that time and are now in common use for the extraction of aquatic humic substances (Aiken, this volume).

Although we are unaware of new resins with more efficiency for sorption of humic substances than those described by Aiken (this volume), there are a number of properties that would be useful to incorporate in a new resin. They include low resin bleed, high capacity for humic substances, efficient elution, low retention of inorganic substances, and regeneration capability.

Although none of the following resins have all these properties, there are other resins available for consideration. They include silica and alumina liquid chromatographic packings, preparative C-18 chromatographic packings, and PRP-1 (styrene divinylbenzene resin from Hamilton, Salt Lake City, Utah, U.S.A.). New resins that may be considered for future research include synthetic zeolites, Al-C_{18} reverse phase, and composite resins (e.g., Si polyurethane).

Fractionation

In the area of fractionation of *aquatic* humic substances, certain properties of the humic material should be kept in mind and exploited for separation, including:

1) metal binding (ligand exchange chromatography);
2) hydrophilic/hydrophobic interaction as a function of pH (reverse phase chromatography);
3) macromolecular properties (permeation chromatography);
4) acidic carboxyl and phenolic hydroxyl groups (weak base ion-exchange chromatography);
5) pi electon interactions (cyano-propyl resins); and
6) nitrogen content (cation-exchange chromatography).

All of these factors should be considered in future chromatographic approaches used to separate and fractionate humic substances. Also, the methods listed above for separation are applicable for *aquatic* humic substances.

General considerations

There is a general need for a collection of humic and fulvic samples from a variety of different environments in soil and water (MacCarthy 1976). The IHSS is providing a service to its members through banking of both standard and reference samples of humic substances. The purpose of this bank is to provide material for quality control of analytical work on humic substances, to allow different laboratories to compare results for identical material, and to build up a large reservoir of analytical and other data for the benefit of all workers.

Table 1 Several of the standard and reference samples of the IHSS.

Sample	Location
Aquatic humic and fulvic acids	Suwannee River, Florida, U.S.A.
Mollisol humic and fulvic acids	Illinois, U.S.A.
Leonardite humic acids	Wyoming, U.S.A.
Peat humic and fulvic acids	U.S.A.

Standard and reference samples are both available from the same origin. However, the samples have been split and are available for two different purposes: standard sample (100 mg amounts only) and reference sample (larger amounts for research use).

These materials are banked and available for purchase from the secretarial office of the IHSS (Dr. Patrick MacCarthy, Dept. of Chemistry, Colorado School of Mines, Golden, Colorado 80401, U.S.A.). Scientists working on humic substances from different environments are encouraged to donate their own humic material to this bank so that future comparative research may be done. Table 1 shows the samples available at the present time. Care must be taken in the use of commercial humic substances because they commonly do not represent humic substances from soil and water (Malcolm and MacCarthy 1986).

Finally, any isolate of *soil* and *aquatic* humic substances is intrinsically polydisperse and any chemical properties derived from these substances are average values. Therefore, we recommend that the property of polydispersity be addressed in all humic and fulvic isolates in order to guide our average and range values.

SUMMARY

There are several key points that were addressed during the session of the isolation group, the most important being the reconciliation of terms among researchers in aquatic and soil humic substances and the definition of a general flow scheme and nomenclature for the terms humic and fulvic acids. We consider this the major result of our workshop. Second, we consider the use of environmental source descriptors, such as *soil, aquatic,* and *sedimentary* humic substances, essential terms that transmit important information to the reader who is involved in humic research. We recognize that the systems we are studying are complex, but we nonetheless wish to define the operational fractions of humic and fulvic acids. Finally, we realize that our operational definition has excluded discussion of the humification process per se.

RECOMMENDATIONS FOR FUTURE RESEARCH

Recommendations for future research include:

1) Continue research on whole soil and water analysis using ^{13}C NMR and other solid state chemical techniques.
2) Bank humic and fulvic acids with the IHSS for future studies of humic substances and encourage the widespread use of standard and reference samples.
3) Continue research on humic substances from unique aquatic environments in order to understand autochthonous input of humic substances. We realize that soil scientists have essentially completed this task, but it has not been done by those working on *aquatic* humic substances.
4) Begin comparative studies of humic isolation methods and storage properties of humic substances (e.g., storage as sodium salts).
5) Contine efforts on purification of both *aquatic* and *soil* humic and fulvic acids.
6) Make efforts to synthesize adsorbents capable of specific separation of humic substances based on key chemical and physical properties.
7) Increase the research efforts on marine humic substances in sedimentary, particulate and dissolved phases. Finally, continue the much needed dialogue and cooperative work among soil, marine and freshwater chemists working on humic substances.

Acknowledgement.   Special thanks to P. Winchester for editing and typing the manuscript and supporting the discussion group.

REFERENCES

Achard, F.K. 1786. Chemische Untersuchung des Torfs. Crell's *Chem. Ann.* **2:** 391–403.
Aiken, G.R. 1985. Isolation and concentration techniques for aquatic humic substances. In: Humic Substances in Soil, Sediment and Water, eds. G.R. Aiken et al. New York: Wiley-Interscience.
Aiken, G.R.; McKnight, D.M.; Wershaw, R.L.; and MacCarthy, P. 1985. Humic Substances in Soil, Sediment and Water. New York: Wiley-Interscience.
Aiken, G.R.; Thurman, E.M.; Malcolm, R.L.; and Walton, H.F. 1979. Comparison of XAD macroporous resins for the concentration of fulvic acid from aqueous solution. *Analyt. Chem.* **51:** 1799–1803.
Berzelius, J.J. 1806. Afhandlingar i Physik, Kemi, och Mineralogi 1: 124–145.
Brandt, K. 1899. Über den Stoffwechsel im Meere. Wiss. Meeresunters., vol. 4, pp. 213–230. Kiel: Abt.
Ertel, J.R., and Hedges, J.I. 1984. The lignin component of humic substances: distribution among soil and sedimentary humic, fulvic, and base-insoluble fractions. *Geochim. Cosmochim. Acta* **48:** 2065–2074.

Fogg, G.E. 1966. The extracellular product of algae. *Oceanogr. Mar. Biol. Ann. Rev.* **4:** 195–212.

Gjessing, E.T. 1976. Physical and Chemical Characteristics of Aquatic Humus. Ann Arbor, MI: Ann Arbor Science.

Harvey, G.R.; Boran, D.A.; Chesal, L.A.; and Tokar, J.M. 1983. The structure of marine fulvic and humic acids. *Marine Chem.* **12:** 119–132.

Hayes, M.H.B. 1985. Extraction of humic substances from soil. In: Humic Substances in Soil, Sediment and Water, eds. G.R. Aiken et al. New York: Wiley-Interscience.

Kalle, K. 1937. Meereskundliche chemische Untersuchen mit Hilfe des Zeisschen Pulfrich Photometers. *Ann. Hydrogr. Berl.* **65:** 276–282.

Kalle, K. 1966. The problem of Gelbstoff in the sea. In: *Oceanogr. Mar. Biol. Ann. Rev.,* No. 4, ed. Barnes, pp. 91–104. London: Allen and Unwin.

MacCarthy, P. 1976. A proposal to establish a reference collection of humic materials for interlaboratory comparisons. *Geoderma* **16:** 179–181.

Malcolm, R.L., and MacCarthy, P. 1986. Limitations in the use of commercial humic acid in water and soil research. *Env. Sci. Tech.* **20:** 904–911.

Mantoura, R.F.C., and Riley, J.P. 1975. The analytical concentration of humic substances from natural waters. *Analyt. Chim. Acta* **76:** 97–106.

Nissenbaum, A., and Kaplan, I.R. 1972. Chemical and isotopic evidence for the in situ origin of marine humic substances. *Limnol. Oceanogr.* **17:** 570–582.

Pocklington, R. 1977. Chemical processes and interactions involving marine organic matter. *Marine Chem.* **5:** 479–496.

Riley, J.P., and Taylor, D. 1969. The analytical concentration of traces of dissolved organic materials from seawater with Amberlite XAD-1. *Analyt. Chim. Acta* **46:** 307–309.

Schnitzer, M., and Khan, S.U. 1972. Humic Substances in the Environment. New York: Marcel Dekker.

Shapiro, J. 1957. Chemical and biological studies on the yellow organic acids of lake water. *Limnol. Oceanogr.* **2:** 161–179.

Stevenson, F.J. 1982. Humus Chemistry. New York: John Wiley & Sons.

Thurman, E.M. 1985. Organic Geochemistry of Natural Waters. Dordrecht: Martinus Nyhoff.

Thurman, E.M., and Malcolm, R.L. 1981. Preparative isolation of aquatic humic substances. *Env. Sci. Tech.* **15:** 463–466.

Waksman, S.A. 1938. Humus-Origin, Chemical Composition and Importance in Nature, 2nd rev. Baltimore: Williams and Wilkins.

Wilson, M. 1987. NMR Techniques and Applications in Geochemistry and Soil Chemistry. Oxford: Pergamon Press.

Wilson, M.A.; Philip, R.P.; Gillam, A.H.; Gilbert, T.D.; and Tate, K.R. 1983. Comparison of the structures of humic substances from aquatic and terrestrial sources by pyrolysis gas chromatography-mass spectrometry. *Geochim. Cosmochim. Acta* **47:** 497–502.

Humic Substances and Their Role in the Environment
eds. F.H. Frimmel and R.F. Christman, pp. 45–58
John Wiley & Sons Limited
© S. Bernhard, Dahlem Konferenzen, 1988.

Polymerization of Humic Substances in Natural Environments

J.I. Hedges

School of Oceanography, WB–10
University of Washington
Seattle, WA 98195, U.S.A.

Abstract. Although it is generally accepted that humic substances are polymers, the mechanisms by which these large organic molecules are formed remain unclear. This paper reviews the definitive characteristics of humic substances and critically evaluates the role of polymerization in the formation of humic substances by degradation and condensation pathways. Finally, research directions are recommended which might more effectively advance our understanding of the genesis of humic substances in natural environments.

INTRODUCTION

The scope of this background paper is to provide an overview of current thought concerning the role of polymerization in the formation of humic substances in natural environments. "Polymerization" will be used here in reference to the construction of large organic molecules from smaller subunits, although the term technically is restricted only to circumstances where a limited variety of building blocks is involved. The discussion will emphasize conditions within aquatic environments which affect the formation of humic substances and their chemical characteristics. This overview is not comprehensive and represents a personal aquatic perspective on this complex and often controversial topic. Although I have attempted to minimize overlap with other background papers, the subjects of selective degradation, model systems, and isolation all are related to the polymerization question and will be treated briefly here when necessary for completeness or contrast.

DEFINITIVE CHARACTERISTICS OF HUMIC SUBSTANCES

Before looking in detail at various polymerization models, it is informative first to consider the characteristics of humic substances in general and to examine those characteristics which have come to shade our perceptions of their genesis.

One of the most definitive characteristics of humic substances is that these acidic organic molecules are large (molecular weight = 500 to >250,000 Daltons). Since most common biochemical structural units such as simple sugars, amino acids, fatty acids, and cinnamyl alcohols (the building blocks of lignins) have molecular weights less than 500 Daltons, it is generally inferred that humic substances must ultimately result from polymerization reactions involving such small precursors. It is still a matter of debate, however, whether this polymerization occurs metabolically to form biopolymers, which are then marginally altered in the environment to humic substances, or takes place in the later stages of biopolymer decomposition as small reactive degradation products spontaneously recombine. Thus both the site (e.g., intracellular vs. extracellular) and mechanism (e.g., enzymatically mediated vs. spontaneous) of polymerization remain in question.

Probably the second most important characteristic of humic substances in terms of their formation mechanisms is that only a small fraction (10–20%) of the polymers typically can be accounted for as recognizable biochemicals (Thurman 1985). Although a comprehensive inventory of biochemicals is difficult to establish even for living organisms, this low percentage has led to the general assumption that humic substances must be the product of extensive microbial degradation which has altered much of the initial biochemical signature. Spontaneous (abiotic) polymerization reactions are thought by many to further attenuate the pool of recognizable biochemicals within humic polymers as the simple structural units are modified in the course of condensation.

Another universal property of humic substances is that they contain significant amounts of nitrogen. Amino acids, the most common nitrogenous substances in living organisms, however, exhibit little tendency to spontaneously combine with each other under most "natural" conditions (Hedges 1978) and are themselves usually a minor component of humic polymers (Thurman 1985). Therefore, the presence of nitrogen in humic substances has been taken as evidence for polymerization reactions which involve nitrogenous precursors. Microbial degradation is implicated in this scheme as a means of introducing the needed nitrogenous reactants, either in the form of simple biochemicals or ammonia.

Humic substances are typically yellow to brown in color and acidic. Thus, humic molecules must contain, in addition to acidic functional groups, conjugated systems of resonating electrons capable of absorbing blue light.

Although the acidity of humic substances can be assumed to be largely due to the presence of carboxyl and phenolic functionalities, the range of possible chromophores is broad and includes both benzene and furan ring systems.

The final definitive characteristic of humic substances is their ubiquity. Their universal distribution indicates either that many precursors and formation pathways exist or that conditions favoring a limited number of formation mechanisms are widely found in nature. Distinguishing these two alternate explanations has been difficult because individual scientists have tended to concentrate their efforts on specific types of samples (e.g., soil, sediment, or water) for which different isolation methods often are used (Thurman 1985).

SPECIFIC FORMATION MODELS

A wide variety of formation models has been suggested to explain the previously discussed characteristics of humic substances found in different natural environments (Stevenson 1982). One means of catagorizing the various theories is on the basis of whether the reactions *directly responsible* for humic substance formation are assumed to have a net degradative (lowered molecular weight) or aggregative (increased molecular weight) effect (Fig. 1). One of the two classes of models, which will be referred to here as the biopolymer degradation (BD) type, assumes that polymerization occurs within cells via secondary metabolism to form biopolymers that are then partially degraded in the environment to produce humic substances (Fig. 1). This model type, in which the integrity of the original biopolymer structure is not completely destroyed by microbial degradation, is discussed in detail by Hatcher and Spiker (this volume).

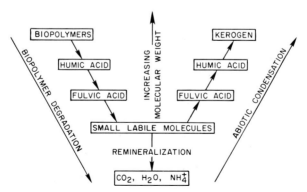

Fig. 1—Schematic representation of biopolymer degradation (BD) and abiotic condensation (AC) for humic substance formation.

The other class of models involves the condensation, or "repolymeri-zation," of reactive small organic molecules that have been generated by essentially complete breakdown of bonds between structural units in the original biopolymers (Fig. 1). Such representations will be referred to here-after as abiotic condensation (AC) models, although it is clear that the precursor substances ultimately are biologically formed and enzymatically degraded. Because of their constructive latter phase, AC models usually assume fulvic acids to be a precursor of humic acids, whereas the opposite is required for BD models.

It is useful to note that the two reaction pathways envisioned in the BD and AC models are not mutually exclusive and in fact biopolymer degradation is a step in the AC pathway. The key questions involved in distinguishing these two model categories are (a) whether slow erosional degradation of large biopolymers can directly form humic substances and (b) whether the degra-dation products of biopolymers spontaneously recombine to a significant extent in nature. Of course, the choice to envision either pathway as an orderly sequence with well defined precursor-product relationships (Fig. 1) is probably a serious oversimplification. For example, both small and large molecules can be simultaneously released (without subsequent interconver-sion) from a degrading biopolymer and the formation or cleavage of only one chemical bond can reverse molecular weight trends within either the BD or AC pathways.

Biopolymer degradation models

Since biopolymer degradation is a common step in both proposed model types, it is most reasonable to begin with a general description of this proposed pathway for humic substance formation (see Hatcher and Spiker, this volume, for a detailed discussion). In general, the greatest appeal of BD models is their assumption that the framework of the humic polymers is assembled by living organisms via enzymatic mediation. In this way large molecules can be constructed without requiring the high concentrations or reactivities of precursors needed for extracellular polymerization. In addition, if the resulting polymer is relatively refractory toward microbial disassembly, it then has a high probability to persist in natural environments where it can trap (or "immobilize") reactive nitrogen compounds. Thus only the nitrogenous reactant need be labile toward microbial decay. The dual requirements for (a) stable bonds to retain the integrity of the original polymer as well as (b) reactive subunits to immobilize nitrogen would be best met by a structually complex biopolymer with a variety of linkages between and within structural units.

Waksman's (1932) "lignin-protein" scheme for the formation of soil humic substances via the partial degradation of lignin polymers probably has been the most widely accepted BD model. Although other BD models have not received as much attention, it is possible that other refractory biopolymers, possibly including polysaccharides (Martin and Haider 1971), have not as yet been recognized. In addition, microbial alteration in some cases may mask known biopolymers without causing extensive degradation. The possibility that unknown or unrecognizable biopolymers might occur in humic substances is greater than might be thought because the usual preservation of 1% or less of plant primary production (Abelson 1978) allows for concentration factors of over two orders of magnitude for minor refractory components. In addition, the properties of high molecular weight and chemical stability that would be expected for effective biopolymer precursors would make such substances difficult to detect and characterize by conventional degradative chemical analyses. The recent report of a novel highly aliphatic biopolymer in plant cuticles (Nip et al. 1986) is a good demonstration of the possible existence of a wider range of refractory biopolymer precursors.

Abiotic condensation models

Models of the AC type, on which this overview focuses, are presently the most popular theories for humic substance formation in marine and terrestrial environments. As previously discussed, these models can explain both the polymeric nature of humic substances and the dearth of identifiable biochemical components. They also can account for the high apparent structural diversity of the polymers by randomly building them up from a complex mixture of precursors. As a class, however, AC models are also more difficult to test than their BD counterparts, because there is a less well defined starting point and mechanism for formation. This may contribute to the popularity and persistence of this class of models.

   The polyphenol model, which, among others, was popularized by Kononova (1966), Flaig (1964), and Martin and Haider (1971), is distinct from Waksman's (1932) lignin-protein model in two fundamental ways (Stevenson 1982). First, a variety of phenolic biochemicals other than lignin are assumed to be important precursors and second, the immediate building blocks of humic polymers are thought to be simple phenols as opposed to partially degraded lignin polymer skeletons. Both models, however, assume Schiff base formation between quinones and amines and/or ammonia to be an important mechanism for both polymerization and the inclusion of nitrogen.

   One of the major strengths of the polyphenol model is that the polymerization of quinones is an extremely facile reaction under environmental conditions, especially at slightly basic pH or in the presence of nitrogenous substances (Stevenson 1982). The proposed reaction can easily be simulated in the laboratory with simple quinones or quinones produced by the oxidation of *ortho* or *para* polyphenols with agents such as AgO or $K_2S_2O_8$. The quinones readily condense with each other, or with amino acids and ammonia, to form synthetic polymers which have bulk chemical properties (Ladd and Butler 1966; Ertel and Hedges 1983) and $KMnO_4$ degradation products (Mathur and Schnitzer 1978) that closely resemble those of natural soil humic substances. In natural environments the oxidation of polyphenols to quinones can either occur spontaneously in the presence of molecular oxygen or be enzymatically mediated by a wide variety of microorganisms (Stevenson 1982). Such abiotic oxidation reactions are reportedly catalyzed by clay minerals (Wang et al. 1978), insoluble transition metal oxides, and dissolved cations such as Mn^{2+} and Fe^{3+} (Larson and Hufnal 1980).

An additional strong point of this AC model is that simple phenols with the required substitution patterns to make quinones are known to be common microbial degradation products of lignin (Kirk 1984) and occur in other biochemicals such as tannins. In addition, such phenols are synthesized by fungi of the *Imperfecti* group (Martin and Haider 1971), some species of *streptomycetes* and marine kelps (Ragan 1978). Direct production of dark-colored, humic-like polymers from phenolic precursors is well documented for soil fungi of the *Imperfecti* group that were cultured on nonaromatic substrates (Martin and Haider 1971). Thus, unlike the lignin-protein model, the polyphenol pathway is not necessarily restricted to terrigenous environments or extracellular reactions.

Finally, since many microorganisms have difficulty degrading aromatic substances, phenols, have a reasonably large probability of persisting at high enough concentrations to be effective quinone precursors. The microbial degradation of monophenols and unoxygenated aromatic compounds can also potentially lead to quinone formation because these biochemical pathways involve vicinal dihydroxy intermediates. In addition, many fungi secrete oxidative exoenzymes capable of forming quinone intermediates from lignins and other phenolic precursors (Stevenson 1982). The ability of many polyphenols to combine with and inhibit the microbial degradation of proteinaceous substances has long been recognized (e.g., in tanning) and could help explain the persistence of these nitrogenous polymers in natural environments.

The polyphenol model, however, does have some drawbacks, especially as an important pathway for the production of marine humic substances. One problem is that, although present, phenols typically are not abundant in phytoplankton, which are the major producers of organic material in aquatic environments. Thus the probability of bringing together reactive phenols dissolved in seawater and sediment pore waters is apt to be lower than in terrigenous environments where lignin and fungi are more abundant. A second problem is that dissolved and sedimentary marine humic substances typically do not exhibit the high inherited aromaticity (H/C < 0.75) of synthetic phenol-derived polymers as indicated both by elemental analysis (Ertel and Hedges 1983) and ^{13}C NMR (Gillam and Wilson 1986). Finally, the oxidation of catechols to quinones is difficult under the suboxic conditions encountered beneath the very surface of most coastal marine sediments.

The melanoidin or "browning reaction" model, which involves condensation reactions between simple sugars and amino acids, was first introduced by Maillard (1913) and has been periodically studied ever since (e.g., Hoering 1973; Ikan et al. 1986). The reaction is initiated by formation of a Schiff base between the carbonyl of a sugar and the nitrogen of an amino acid or ammonia. The resulting N-substituted glucosamine then undergoes a complex series of dehydration, rearrangement, and condensation reactions to produce both simple fragmentation products and structurally complex brown nitrogenous polymers (Stevenson 1982). The amino acid moiety is typically altered in the early stages of the reaction. In the absence of nitrogenous material, simple sugars will condense at almost the same rate with each other to form dark acidic polymers (Hedges 1978).

The appeal of the melanoidin model, especially for marine systems, is that the two proposed precursors for humic substance formation are among the most abundant constituents of all living organisms and will combine under reducing as well as oxidizing conditions. In addition, the condensation of reducing sugars with basic amino acids is kinetically favored over reactions with other amino acid types, leading to nitrogen-rich polymers which resemble marine humic substances in many of their bulk chemical properties (Hoering 1973; Hedges 1978; Rubinsztain et al. 1984; Ishiwatari et al. 1986). These similarities also are evident in solid-state ^{13}C NMR spectra and may be related to the presence of a high density of furan structural units (Ikan et al. 1986). Nitrogen-rich synthetic melanoidins are efficiently adsorbed to the edges and interbasal surfaces of clay minerals, thereby providing a possible mechanism for their concentration and preservation in marine sediments (Hedges 1978).

The melanoidin hypothesis, however, also has upon closer inspection major problems. Probably the greatest of these is that this second-order reaction is not rapid at room termperature (Hedges 1978). Thus significant rates of condensation of dissolved free sugars and amino acids at the part per billion concentrations typical of most aquatic systems become highly unlikely under natural conditions. This is a particular problem for free sugars and amino acids, which have mean turnover times with respect to microbial degradation on the order of minutes in many aquatic systems and therefore are unlikely to persist long enough to condense with each other. In addition, the spectral match of melanoidins and natural humic substances is less than perfect. In particular, melanoidins typically exhibit strong ^{13}C NMR absorbances for carbonyl carbon near 200 ppm (Ikan et al. 1986; Hedges et al., in preparation), which are rarely observed in natural marine humic

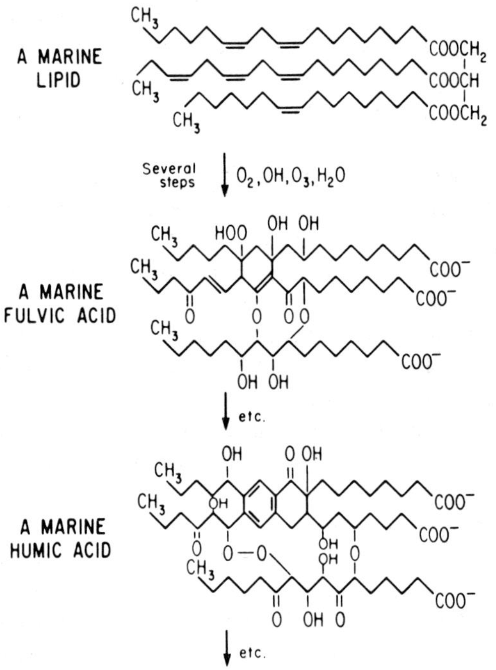

Fig. 2—Representative structures of precursors and products in the photooxidative formation of aquatic humic substances from polyunsaturated lipids (figure directly from Harvey and Boran 1985).

substances (Hatcher and Orem 1986; Gillam and Wilson 1986). Conclusive molecular-level evidence that melanoidins exist naturally has yet to be presented and will be a challenge to provide because most of the degradative reactions commonly used for structural analysis of such polymers can themselves form melanoidin-like material from polysaccharide and protein precursors as a reaction side-product.

The polyunsaturated lipid model for humic substances formation (Fig. 2) involves autooxidative crosslinking reactions between adjacent polyunsaturated fatty acids in fat molecules exposed to sunlight at the ocean surface (Harvey et al. 1983). Aromatic chromophores and additional oxygen-containing functional groups are thought to be introduced in the course of photolysis. This mechanism has been presented specifically to explain the presence of relatively aliphatic, oxygen-rich humic substances in seawater.

As presented, the polyunsaturated lipid model has a number of appealing characteristics. A particular weakness of any AC-type model for the *in situ* formation of aquatic humic substances is that two molecules must condense at a rate that is proportional to the square of their aqueous concentrations.

In most natural waters, and seawater in particular, organic concentrations are on the order of parts per million to parts per billion, making simple condensation reactions highly improbable singular events. The polyunsaturated lipid model circumvents this problem by concentrating the reactive carbon double bonds within a few Angstroms of each other in the three closely packed hydrophobic tails of dissolved triacylglycerols where the unsaturated linkages almost always occur at adjacent points at the fatty acid chains. Under these conditions oxidative crosslinking is potentially rapid and could serve to hold the polymer together even if the labile glycerol ester links are subsequently hydrolyzed (Harvey et al. 1983). Moreover, polyunsaturated fats are commonly found in many types of marine phytoplankton and, due to their surface activity, could be expected to concentrate at the sea surface where ultraviolet light is most intense. Experiments with added polyunsaturated fats reportedly produce humic type condensation products which resemble dissolved humic substances isolated from seawater (Harvey and Boran 1985).

Despite its appealing characteristics, the polyunsaturated lipid model has some problems. One of the main incongruities with the reported characteristics of marine humic substances concerns stable carbon isotope composition. The lipds of marine plankton typically have $^{13}C/^{12}C$ ratios which are decidedly lower than the stable carbon isotope ratios in the total organism. Since the distribution of stable carbon isotopes within the hydrocarbon chains of fatty acids uniformly alternate (Monson and Hayes 1982), the envisioned conversion of polyunsaturated lipids to humic substances (Fig. 2) should occur without appreciable isotope fractionation. However, the stable carbon isotope compositions that have been reported for total dissolved organic carbon (DOC) or humic substances isolated from seawater more closely resemble those measured in the nonlipid components of local marine plankton (Stuermer and Harvey 1974; Meyers-Schulte and Hedges 1986). Although Harvey and Boran (1985) report a comparable fractionation in the phytochemical production of synthetic humic substances from polyunsaturated lipids of known initial stable carbon isotope composition, the incompatibility of this result with theory suggests the need for additional testing.

Another problem is that the proposed model does not include a clear mechanism for the incorporation of nitrogen into the lipid-derived photolysis products (Harvey and Boran 1985). Organic amines, which are the most likely precursors of the nitrogen in marine humic substances, will carry a positive charge at the pH of essentially all natural waters. These functionalities, therefore, would be among the least likely to be found within the hydrophobic marine microenvironments where light-induced lipid crosslinking should occur. However, definitive tests of formation models based on nitrogen content must await resolution of the present discrepancy between the atomic C/N ratios of about 8, published for dissolved humic substances eluted from XAD-2 columns with NH_4OH (Stuermer and Harvey 1974), and ratios near 35 for humics eluted with NaOH (Meyers-Schulte and Hedges 1986). Another unanswered question concerning this model is what importance the proposed mechanism might have for the formation of sedimentary humic substances.

POTENTIALLY PRODUCTIVE RESEARCH DIRECTIONS

The question of future research directions for a better understanding of the role of polymerization in humic substance formation is a difficult one; no one person recognizes all the problems, let alone the solutions. There are some lines of research, however, which seem to offer promise.

Improve and test isolation procedures

The problem of developing suitable isolation procedures is particularly acute for marine humic substances. One such problem is the direct extraction of biochemicals from plant debris in the course of conventional humic substance isolations from sediments. For example, it is clear from model laboratory studies that lignins, carbohydrates, and nitrogenous substances all can be directly recovered in high yields from both fresh and partially degraded vascular plant tissues that have been subjected to standard humic acid and fulvic acid isolation procedures (Stevenson 1982; Ertel and Hedges 1985). There is little reason to expect that the same would not be true of other organisms including marine plankton. Unfortunately, it is essentially impossible to completely remove discrete particles of organic matter from a sediment sample prior to base extraction. It is also difficult to distinguish biochemically recognizable structural units that preexisted as a part of sedimentary humic coatings on mineral grains from those which were independently isolated from physically separate plant fragments in a very different state of preservation. This problem is particularly severe for lake and coastal marine sediments where much of the bulk terrigenous organic matter appears to exist in vascular plant fragments (Ertel and Hedges 1985) as opposed to translocated humic substances such as are found dissolved in water or immobilized within the B1 horizons of soils.

In addition to the previously mentioned problem of the direct extraction of plant tissue remains, isolations of humic substances from marine sediments are particularly susceptible to side reactions with reduced inorganic species such as ammonia and hydrogen sulfide. In light of evidence for facile reactions of these reduced chemical species with a wide variety of biochemicals (e.g., Mango 1983) and humic substances themselves (e.g. Francois 1987), the significance of elemental compositions determined for sedimentary marine humic substances should be questioned. Ways of removing reactive inorganic chemicals prior to humic substance extraction should be investigated.

The high potential for artifact formation in the isolation of dissolved humic substances from seawater can be appreciated by considering that in most techniques milligram quantities of humic material are adsorbed at highly elevated concentrations onto kilogram quantities of resin. All materials are subsequently exposed to a transition of over 10 pH units, often in the presence of reactive nitrogen compounds (Thurman 1985). To compound problems, recoveries are typically inefficient (<10%) and almost certainly result in the isolation of atypically hydrophobic material which is not necessarily polymeric (Thurman 1985). We need to learn new isolation techniques as well as charaterization techniques that might be applied to bulk water samples.

Molecular level characterizations

One way to discriminate between BD and AC formation pathways is to focus molecular level analyses toward the larger, more stable structural units in natural humic substances. For example, the Waksman and the polyphenol models might be discriminated by characterizing intact aromatic dimers and trimers that might be related either to lignin or to preferred condensation products of quinones. Such a study would necessitate the development of degradation reactions, which are relatively free of side reactions, and the use of analytical methods, such as supercritical HPLC, which are applicable to large molecules. In the case of the melanoidin hypothesis, unambiguous molecular indicators of *preexisting* browning reaction products are needed (possibly furans or nitrogen heterocycles). A method for discriminating between furan and benzene carbon double bonds might also be extremely helpful in distinguishing the melanoidin and phenol formation pathways.

Bulk chemical characterizations

Additional means of characterizing intact humic substances (such as cross-polarization/magic-angle spinning ^{13}C NMR) are needed, owing to the absence of clean degradative procedures. In view of the key role that nitrogen apparently plays in humic substance formation, ^{15}N NMR has immense potential in model studies where artifically enriched nitrogen compounds can be used. Characterizations of unfractionated humic substances by soft ion bombardment mass spectrometry may also be useful. The recent development of accelerator techniques for directly counting milligram quantities of carbon offers great potential for testing theories of humic substance formation which involve a specific precursor-product relationship between humic and fulvic acids or their structural constituents (Hedges et al. 1986).

Along these same lines, field evidence that might be used to discriminate constructive from destructive pathways of humic substance formation is particularly slim for aquatic environments. The present general inclination toward AC models by aquatic chemists can be traced largely to the classic publications of Nissenbaum et al. (1971) and Brown et al. (1972), which suggested a progressive conversion of biochemicals to geopolymers in the reducing pore waters and sediments of Saanich Inlet. This evidence was subsequently supported by the work of Krom and Sholkovitz (1977) who demonstrated a preferential increase with depth in the high molecular weight fraction of dissolved organic matter in the pore waters of another reducing marine sediment. Surprisingly few measurements have subsequently been made to test this model using samples from other natural aquatic environments. In particular, the critical distinction of whether fulvic acids are the precursors (AC models) or products (BD models) of humic acids in various natural environments remains to be made.

There probably will never be one key experiment which clearly unlocks the secret of the origin of natural humic substances. However, the current rapid development of new analytical methods and the recent introduction of a number of novel formation models both bode well for this field of research. In spite of the complexities and pitfalls, these are exciting times to be studying the problem of the origin of humic substances.

Acknowledgments.   The preparation of this manuscript was supported by National Science Foundation Grant OCE–8421023. This paper benefitted from reviews by K. Weliky, J. Ertel, and P. Hatcher.

REFERENCES

Abelson, P.H. 1978. Organic matter in the earth's crust. *Ann. Rev. Earth Planet. Sci.* **6**: 325–351.

Brown, F.S.; Baedecker, M.J.; Nissenbaum, A.; and Kaplan, I.R. 1972. Early diagenesis in Saanich Inlet, a reducing fjord, British Columbia. Part III: Early changes in the organic matter. *Geochim. Cosmochim. Acta* **36**: 1185–1203.

Ertel, J.R., and Hedges, J.I. 1983. Bulk chemical and spectroscopic properties of marine and terrestrial humic acids, melanoidins and catechol-based synthetic polymers. In: Aquatic and Terrestrial Humic Materials, eds. R.F. Christman and E.T., Gjessing, pp. 143–163. Ann Arbor: Ann Arbor Science.

Ertel, J.R., and Hedges, J.I. 1985. Sources of sedimentary humic substances: vascular plant debris. *Geochim. Cosmochim. Acta* **49**: 2097–2107.

Flaig, W. 1964. Effects of microorganisms in the transformation of lignin to humic substances. *Geochim. Cosmochim. Acta* **28**: 1523–1535.

Francois, R. 1987. A study of the effect of extraction conditions on the artifactual inclusion of sulfur in humic acids. *Geochim. Cosmochim. Acta* **51**: 17–27.

Gillam, A.H., and Wilson, M.A. 1986. Structural analysis of aquatic humic substances by NMR spectroscopy. In: Organic Marine Geochemistry, ed. M.L. Sohn, pp. 128–141. Washington, D.C.: American Chemical Society.

Harvey, G.R., and Boran, D.A. 1985. Geochemistry of humic substances in seawater. In: Humic Substances in Soil, Sediment and Water, eds. G.R. Aiken, D.M. McKnight, R.L. Wershaw, and P. MacCarthy, pp. 233–247. New York: John Wiley and Sons.

Harvey, G.R.; Boran, D.A.; Chesal, L.A.; and Tokar, J.M. 1983. The structure of marine fulvic and humic acids. *Marine Chem.* **12**: 119–132.

Hatcher, P.G., and Orem, W.H. 1986. Structural interrelationships among humic substances in marine and estuarine sediments as delineated by cross-polarization/magic angle spinning ^{13}C NMR. In: Organic Marine Geochemistry, ed. M.L. Sohn, pp. 142–157. Washington, D.C.: American Chemical Society.

Hedges, J.I. 1978. The formation and clay mineral reactions of melanoidins. *Geochim. Cosmochim. Acta* **42**: 69–76.

Hedges, J.I.; Ertel, J.R.; Quay, P.D.; Grootes, P.M.; Richey, J.E.; Devol, A.H.; Farwell, G.W.; Schmidt, F.W.; and Salati, E. 1986. Organic carbon-14 in the Amazon River system. *Science* **231**: 1129–1131.

Hoering, T.C. 1973. A comparison of melanoidin and humic acid. *Carnegie Inst. Wash. Year Book* **72**: 682–690.

Ikan, R.; Rubinsztain, Y.; Ioselis, P.; Aizenshtat, Z.; Pugmire, R.; Anderson, L.L.; and Woolfenden, W.R. 1986. Carbon-13 cross polarized magic-angle samples spinning nuclear magnetic resonance of melanoidins. *Org. Geochem.* **9**: 199–212.

Ishiwatari, R.; Morinaga, S.; Yamamoto, S.; Machihara, T.; Rubinsztain, Y.; Ioselis, P.; Aizenshtat, Z.; and Ikan, R. 1986. A study of formation mechanism of sedimentary humic substances. I. Characterization of synthetic humic substances (melanoidins) by alkaline potassium permanganate oxidation. *Org. Geochem.* **9**: 11–23.

Kirk, T.K. 1984. Degradation of lignin. *Microbiol. Ser.* **13**: 399–437.

Kononova, M.M. 1966. Soil Organic Matter. Oxford: Pergamon.

Krom, M.D., and Sholkovitz, E.R. 1977. Nature and reactions of dissolved organic matter in the interstitial waters of marine sediments. *Geochim. Cosmochim. Acta* **41**: 1565–1573.

Ladd, J.N., and Butler, J.H.A. 1966. Comparisons of some properties of soil humic acids and synthetic phenolic polymers incorporating amino derivatives. *Aust. J. Soil Res.* **4**: 41–54.

Larson, R.A., and Hufnal, J.M. Jr. 1980. Oxidative polymerization of dissolved phenols by soluble and insoluble inorganic species. *Limnol. Oceanogr.* **25**: 505–512.

Maillard, L.C. 1913. Formation de matieres humiques par action de polypeptides sur sucres. *C.R. Acad. Sci.* **156**: 148–149.

Mango, F.D. 1983. The diagenesis of carbohydrates by hydrogen sulfide. *Geochim. Cosmochim. Acta* **47**: 1433–1441.

Martin, J.P., and Haider, K. 1971. Microbial activity in relation to soil humus formation. *Soil Sci.* **111**: 54–63.

Mathur, S.P., and Schnitzer, M. 1978. A chemical and spectroscopic characterization of some synthetic analogues of humic acids. *Soil Sci. Soc. Am. J.* **42**: 591–596.

Meyers-Schulte, K.J., and Hedges, J.I. 1986. Molecular evidence for a terrestrial component of organic matter dissolved in ocean water. *Nature* **321**: 61–63.

Monson, K.D., and Hayes, J.M. 1982. Carbon isotopic fractionation in the biosynthesis of bacterial fatty acids. Ozonolysis of unsaturated fatty acids as a means of determining the intramolecular distribution of carbon isotopes. *Geochim. Cosmochim. Acta* **21**: 110–126.

Nip, M.; Tegelaar, E.W.; de Leeuw, J.W.; Schenck, P.A.; and Holloway, P.J. 1986.

Analysis of modern and fossil plant cuticles by Curie point Py-GC and Curie point Py-GC-MS: recognition of a new, highly aliphatic and resistant biopolymer. *Org. Geochem.* **10**: 769–778.

Nissenbaum, A.; Baedecker, M.J.; and Kaplan, I.R. 1971. Studies on dissolved organic matter from interstitial water of a reducing marine fjord. In: Advances in Organic Geochemistry, pp. 427–440. Oxford: Pergamon Press.

Ragan, M.A. 1978. Phenolic compounds in brown and red algae. In: Handbook of Phycological Methods, vol. II, eds. J.A. Hellebust and J.S. Craigie, pp. 157–179. Cambridge: Cambridge Univ. Press.

Rubinsztain, Y.; Ioselis, P.; Ikan, R.; and Aizenshtat, Z. 1984. Investigations on the structural units of melanoidins. *Org. Geochem.* **6**: 791–804.

Stevenson, F.J. 1982. Humus Chemistry (Genesis, Composition, Reactions). New York: John Wiley-Interscience.

Stuermer, D.H., and Harvey, G.R. 1974. Humic substances from seawater. *Nature* **250**: 480–481.

Thurman, E.M. 1985. Organic Geochemistry of Natural Waters. Boston: W. Junk.

Waksman, S.A. 1932. Humus. Baltimore: Williams and Wilkins.

Wang, T.S.C.; Li, S.W.; and Huang, P.M. 1978. Catalytic polymerization of phenolic compounds by a latosol. *Soil Sci.* **126**: 81–86.

Humic Substances and Their Role in the Environment
eds. F.H. Frimmel and R.F. Christman, pp. 59–74
John Wiley & Sons Limited
© S. Bernhard, Dahlem Konferenzen, 1988.

Selective Degradation of Plant Biomolecules

P.G. Hatcher and E.C. Spiker

*U.S. Geological Survey
Reston, VA 22092, U.S.A.*

Abstract. Although the genesis of humic substances has been debated since the 1800s, the principal scientific views in favor today still range from large molecule degradation to small molecule condensation. In this chapter we review some modern evidence supporting a degradative pathway, though a condensation pathway is also plausible. In the degradative scheme, resistant plant and microbial biopolymers are the precursors from which humic substances evolve. Increasing degradation leads to a progressive evolution of humin, first, then the more soluble humic acids, and finally the most soluble fulvic acids representing the most humified fraction of humic substances.

INTRODUCTION

It is generally agreed that humic substances originate from degradation of plant remains, but there is little consensus on the mechanisms by which humification occurs. All hypotheses proposed for humification can be grouped into two contrasting categories: one involves a degradative pathway where plant biopolymers and possibly some small molecules linked into this polymer during humification are modified by degradation to form the central core of humic substances; the other hypothesis involves a condensation polymerization pathway in which plant biopolymers are first degraded to small molecules which then re-polymerize. The focus of this chapter is to provide an evaluation of the degradative pathway. In so doing we do not mean to imply that condensation processes (Hedges, this volume) are not possible. Rather, in this chapter we are putting the degradation hypothesis to the test of explaining existing data. In many instances both condensation and degradation may be occurring simultaneously because humification is a dynamic process with no unique unidirectional vector.

As early as the mid-to-late 1800s a number of individuals recognized that humus orginated from the decomposition of vegetation, and numerous hypotheses were proposed for humus formation (Waksman 1938). Many proposed an abiotic degradation of plant components for the formation of humus, but it soon became apparent that microorganisms play an essential role in humification (Waksman 1938). Thus two major hypotheses evolved, one in which humification is microbially mediated and the other which suggested an abiogenic origin for humic substances, though the role of microorganisms was deemed essential in providing the raw materials for abiogenic processes. Fischer and Schrader (1921) proposed that lignin was most likely the mother substance for humus, an hypothesis which became well accepted at the time. This work was most convincing because it was based on laboratory and field studies, both of which showed that decomposition of plant remains led to the removal of cellulose and other carbohydrates with selective preservation of a lignin "core" from which humus was thought to evolve.

Though many aspects of the lignin hypothesis for humus formation seemed attractive, it could not be entirely accepted. Owing to the low nitrogen content of lignin the model failed to account for the nitrogen content of humus, and this weakness has been solely responsible for the general lack of acceptance of the lignin hypothesis. Waksman (1938), however, amended the lignin hypothesis by suggesting that proteinaceous materials, primarily from microorganisms, become linked into the primarily lignin-derived humus "core" as the lignin is modified by the action of organisms. These ideas had a tremendous impact on the contemporaneous understanding of humification. The hypothesis was basically a degradative one whereby humic substances were produced as the lignin became modified and incorporated proteinaceous substances. The humus thus formed has a continually changing structure depending on the degree of decomposition, with higher degrees of oxidative degradation producing the most modified lignin products.

Though Waksman's model for humification was, for the most part, accepted, the development of new chemical analytical and radiolabelling techniques for tracing the degradation of plant remains into the formation of humic substances began to erode Waksman's model. The pioneering studies of Flaig (1966), in which ^{14}C-labeled wheat straw was subjected to degradation, led to new ideas on the role of lignin in humification (see Hedges, this volume). Flaig proposed that lignin was oxidatively degraded to simple phenolic monomers that underwent oxidative polymerization to produce humic substances. The work of Martin and Haider (1971) later showed that many of the phenols produced by fungi and other microorganisms could be polymerized into the humic framework. This and Flaig's work were instrumental in rendering Waksman's model obsolete. However, Flaig reiterated several times that the studies on degradation of wheat straw that formed the basis for his model were inconclusive as to whether humic acids were formed by polycondensation of phenols or were formed by oxidative modification of lignin. In essence, Flaig's studies could also be interpreted to indicate a degradative pathway for the formation of humic acids. The body of evidence that model phenolic compounds, all of which could be produced by oxidative degradation of lignin, could be polymerized via quinone intermediates led Flaig (1966) to prefer the polyphenol condensation hypothesis.

It is fair to state that by the early 1980s condensation/polymerization models were widely accepted by most individuals (Stevenson 1982). Recently, however, the polycondensation concept has been questioned as the development of analytical techniques capable of examining both soluble and insoluble macromolecules have allowed us to track the evolution of biomolecules during degradation. Foremost of these techniques are solid-state ^{13}C NMR, analytical pyrolysis, and oxidative degradation/gas chromatography/mass spectrometry.

EVIDENCE FOR A DEGRADATIVE PATHWAY FOR HUMIFICATION

The humification pathway being evaluated here is essentially that depicted in Fig. 1. Vascular and nonvascular plants and microorganisms are composed of a variety of major classes of biochemical compounds, some better defined than others. Some of these complex macromolecules are only now being discovered (e.g., paraffinic macromolecules in algae (Largeau et al. 1984) and in plant cuticles (Nip et al. 1986)). During microbial degradation the labile macromolecules are degraded and lost, while refractory compounds or biopolymers such as lignin, cutins, suberins, N-containing paraffinic macromolecules, melanins, and possibly other unidentified refractory biopolymers are selectively preserved to become a part of what is operationally defined as humin. With increasing degradation, the humin can be oxidized further to produce macromolecules that may have the same molecular weight as humin but are more highly functionalized (e.g., contain a greater proportion of carboxyl, carbonyl, and hydroxyl groups). The increased functionality promotes increased solubility in alkali thereby leading to the evolution of humic acids. Increased degradation yields greater functionality, lower molecular weight, and greater solubility in acid and base (fulvic acids). As a consequence of increased degradation, the macromolecules become more highly modified structurally and begin to lose their chemical similarity to parent material.

In this degradative model, a modification of Waksman's model, fulvic acids are diagenetically downstream of humic acids which are diagenetically downstream of humin, which is essentially a partially modified accumulation

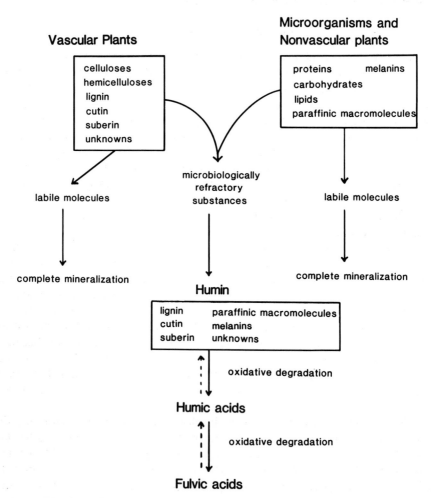

Fig. 1—The degradative pathway for formation of humic substances. Dotted arrows indicate that the transformation is possible but is of minor importance.

of plant biopolymers. This model can be applied to both terrestrial vascular plant remains and to microbial and/or nonvascular plant remains. The new chemical structural information obtained for humic substances is beginning to provide some valuable insights on how these substances originate. It is not the intent in this chapter to discuss the new structural information in detail, but it is important to consider the modern concepts of the chemical composition of humic substances to better evaluate their origins.

Modern concepts of humic substances related to genesis

It is well recognized that humic substances are macromolecular entities having chemical compositions that vary in subtle ways among different environments. The variations noted are subtle only because the analytical methods used are not overly specific and are probably biased in their delineation of the complete structure, e.g., harsh degradative techniques. The combined application of molecular level analysis techniques with spectroscopic structural analysis techniques such as NMR has brought to light some new concepts about the chemical composition of humic substances (Norwood, this volume). It is important to consider modern views of the chemical nature of the various humic isolates, the fulvic acids, humic acids, and humin in order to elaborate on the evidence supporting a degradative pathway for their origin.

Probably the most supportive evidence for degradative pathways for humification comes from close examination of the chemical interrelationships among various humic isolates. Comparison of humic isolates with original plant material is of minor significance because the bulk of fresh plant material is degraded during humification (Waksman 1938). However, minor, resistant fractions of fresh plant material are worthy of comparison with humic substances within the context of a degradative scheme for humification because they may become enriched by large factors during degradation. In the case of vascular plants, the nature of lignin is well understood, but other biopolymers such as the aliphatic materials from cuticles (Nip et al. 1986) are less known as contributors to humification.

Humin, the material that remains insoluble after alkali extraction of sediments, is generally considered to be insoluble because of its large molecular size, though soil scientists believe that it represents a clay-humic acid complex (Stevenson 1982). Stevenson argued that humin in terrestrial sediments should not be considered a humic substance because it has many of the chemical properties of plant biopolymers such as lignin and polysaccharides. Though it is likely that unaltered lignin and polysaccharides are present in humin, the evidence derived from NMR studies (Hatcher et al. 1985) points to the fact that lignin, if present in operationally defined humin, has been modified slightly by degradation. The question becomes what will one accept as being true humin, a modified lignin or some polycondensate that has achieved insolubility by macromolecular size or linkage with clays. Obviously if one is a proponent of the polycondensation hypothesis, the latter alternative is more satisfying; but if one considers a degradative scheme for humification, then humin operationally represents partially altered biopolymers that are intermediates in humification. In this case, the carbohydrate polymers are degraded to soluble monomers to be utilized for microbial synthesis or energy whereas the lignin is more refractory and is only partially altered to become the major fraction of humin.

NMR studies of humin from a variety of depositional environments show that lignin-like structures in humin are best preserved in anaerobic peats, whereas the humin from aerobic soils shows little resemblance to lignin in that characteristic methoxyl and aryl ether or phenolic carbons are absent (Hatcher et al. 1985). Thus, increasing oxidative degradation appears to significantly alter the chemical structure of insoluble humin. The lack of significant aryl ether or phenolic carbons in NMR spectra of humin from aerobic soils is particularly noteworthy considering that phenolic carbons have always been thought of as characteristic of humic substances (Schnitzer and Khan 1972) and that the polyphenol condensation hypothesis is thought to involve linking of phenolic substances (Flaig 1966).

Significant quantities of noncarbohydrate aliphatic structures (paraffinic structures) are present in terrestrial humin from peat and soils (Hatcher et al. 1985). Though the source of these paraffinic structures is currently under debate, it is possible that they arise from algal residues (Hatcher, Spiker et al. 1983) and/or leaf cuticles (Nip et al. 1986). Alternatively, they could arise from degraded lignin, especially if oxidative degradation leads to aromatic ring rupture (Flaig 1966). They could also be derived from protein-derived residues that become linked into the structure of humic substances during humification (Waksman 1938).

In marine sediments or sediments dominated by inputs of nonvascular or non lignin-containing plants, it is difficult to envision the humin as degraded lignin. However, resistant components operationally defined as humin can be recovered in significant quantities in recently deposited material (Hatcher, Spiker et al. 1983; Cronin and Morris 1982). Organic geochemical studies indicate that this humin has mostly an aliphatic structure (paraffinic) and is, thus, chemically unrelated to lignin (Hatcher, Spiker et al. 1983). As in the case of terrestrial humin, the question remains as to whether this paraffinic material is an original biopolymer modified by degradation or is a polycondensate. This material may be viewed as a degraded biopolymer that is resistant but undergoes alteration with increasing degradation, similar to the relationship between terrestrial humin and degraded lignin.

Humin from marine sediments is particularly distinct from terrestrial humin in its nitrogen content and stable carbon isotopic composition (Nissenbaum and Kaplan 1972). The isotopic composition indicates that it must originate from marine algae. The high nitrogen content ($C/N \simeq 10$) is suggestive of proteinaceous materials. It is tempting to suggest that this paraffinic, nitrogen-rich polymer in marine humin is structurally similar to that present in terrestrial humin, given that algal or microbial substances are also present in terrestrial sediments. The presence of paraffinic structures and nitrogen in terrestrial humin might be well explained by contributions from algal or microbial humin. If the nitrogen content in terrestrial humic substances could be explained by incorporation of nonproteinaceous nitrogen-rich algal or microbial residues there would be no need to chemically link proteinaceous materials with lignin degradation products during humification as Waksman (1938) suggested.

Humic acids, the alkali soluble-acid insoluble humic isolates, are often chemically distinct from humin, although the many similarities have led some to genetically link humic acids to humin (Schnitzer and Khan 1972). The similarities based on elemental compositions, infrared and NMR spectra, and chemical degradation products have affirmed a genetic link. However, there are subtle differences between humin and humic acids which are probably significant when considering the exact genetic relationship. It is particularly noteworthy that terrestrial humic acids generally contain less methoxyl carbon than humin (Hatcher et al. 1985), pyrolysis products suggest a more advanced stage of lignin decomposition in humic acids (Saiz-Jimenez et al. 1979), and the carboxyl content of humic acids is higher (Hatcher et al. 1985). These structural relationships are consistent with a degradative scheme in which humic acids are genetically downstream of humin and lignin.

Terrestrial humic acids have always been thought of as carboxyl and phenolic-rich substances (Schnitzer and Khan 1972). Recently, the absolute level of phenolic carbons in humic acids has been questioned (Hatcher et al. 1981) because ^{13}C NMR spectra show little phenolic carbon in some humic acids. Apparently, the humic acids showing the lowest levels of phenolic carbon, if any, are from highly oxidized soils, whereas humic acids from peats show significant levels of phenolic carbons and methoxyl carbons, implying the presence of lignin-like materials (Hatcher, Breger et al. 1983). Pyrolysis (Saiz-Jimenez et al. 1979) and CuO oxidation data (Ertel and Hedges 1984) confirm the presence of lignin-like substances in humic acid isolates but the absolute levels are quite variable. Proponents of the poly-phenol condensation model would argue that lignin-like materials in humic acids originate from the linkage of lignin decompositon monomers into the polyphenolic network during polymerization. Alternatively, the presence of lignin-like materials in humic acids may simply indicate that the alteration of lignin during decomposition is incomplete. Such alteration might involve oxidative cleavage of the propyl side chain to form aromatic ketones, alde-hydes, and acids. Evidence that such processes are active can be found in CuO oxidation studies which indicate increased levels of such oxidation products in humic acids compared to lignin (Ertel and Hedges 1984). In an analogous way as that proposed by Flaig for demethylation of lignin-derived monomers, the methoxyphenols of lignin macromolecules could be demethyl-ated to catechol-like structures without depolymerization of the lignin. Consequently, humic acids originating from humin or lignin via this decompo-sition mechanism would be expected to contain fewer methoxyl groups and increased levels of catechol-like structures. Pyrolysis/gas chromatographic studies (Saiz-Jimenez et al. 1979) confirm the presence of catechol-like structures in humic acids. Furthermore, the catechols are likely to be susceptible to oxidation, inducing either ring rupture to form dicarboxylic acids or inducing the formation of quinones believed to be essential in reactions that link other phenolic and proteinaceous materials (Flaig 1966).

Humic acids from strongly oxidizing environments show ^{13}C NMR spectra containing only aromatic and carboxyl carbons with only a hint of phenolic carbons (Hatcher et al. 1981). Yields of phenolic acids upon permanganate oxidation are low compared to benzenecarboxylic acids (Schnitzer and Calderoni 1985), consistent with the NMR data. So, in these instances, soil humic acids are essentially composed of benzenecarboxylic acids. It is difficult to envision a mechanism of phenol condensation reactions as proposed by Flaig (1966) to produce mostly benzenecarboxylic acids. Oxidative ring rupture of catechols formed by demethylation of lignin, followed by oxidation of the aliphatic product so formed, might be a plausible mechanism (Crawford 1981). If such a process acts on depolymerized lignin monomers, it is likely that the products would be converted to CO_2 and other water soluble molecules. However, if such a process acts on lignin monomers linked to a macromolecular framework, then it is feasible to eventually generate benzenecarboxylic acids attached to the macromolecules. Alternatively, if oxidation of demethylated lignin monomers yields quinones that induce polymerization with phenols and nitrogenous substances, the polymerized products would be expected to contain significant amounts of phenolic units (Flaig 1966). The low phenolic carbon content deduced from NMR spectra contrasts sharply with this hypothesis.

Though ^{13}C NMR data have confirmed the existence of aromatic structures, in some cases almost exclusively, paraffinic structures similar to those in humin can often comprise a significant fraction of terrestrial humic acids. As in the case of humin, the origin of these paraffinic structures is speculative; however, their existence in abundance has led to the suggestion that humic acids originate from polymerization of maleic acid (Anderson and Russell 1976). Alternatively, the data suggest that the paraffinic structures originate from algal or microbial residues and/or humin derived from these residues (Hatcher, Spiker et al. 1983).

In marine sediments and other algal-derived sediments, humic acids are predominantly composed of the above paraffinic structures (Hatcher, Breger et al. 1983), though in some cases polysaccharide-like substances could also be present. The polysaccharides most likely originate from alginic acid polymers. Nitrogen contents of marine humic acids are significantly greater than terrestrial humic acids (Nissenbaum and Kaplan 1972). This high nitrogen content has stimulated the development of models which call for condensation of sugars with amino acids (Nissenbaum and Kaplan 1972). However, such models would be unnecessary if one supposes that the complex nitrogen-rich paraffinic structures were modified original components of algae and other microorganisms. The methoxyl contents and yields of lignin-derived phenols from CuO oxidation are extremely low or negligible (Ertel and Hedges 1984), attesting to the fact that only minute amounts of terrestrial plant carbon becomes incorporated in marine humic acids. It is clear that the major oxygen-containing functional groups in marine or algal humic acids are carboxyl, amide, and alcoholic hydroxyl groups. NMR data also indicate small amounts of ketone or aldehyde groups (Hatcher, Breger et al. 1983), and these are more enriched in humic acids than in humin.

Fulvic acids, base and acid soluble isolates, are generally known to be more oxygen rich and carbon poor than humic acids (Schnitzer and Khan 1972). Much of the increased oxygen manifests itself in carboxyl and alcohol functional groups. Molecular weights of fulvic acids are less than humic acids. Thus, these combined gross properties are consistent with the view that fulvic acids represent structures similar to those of humic acids but with greater hydrophilicity due primarily to the increased level of oxygen-containing functional groups (Schnitzer and Khan 1972).

Recent studies of fulvic acids by NMR methods reveal an extremely varying composition depending on the nature of the environment (Hatcher, Breger et al. 1983). In aerobic soils, fulvic acids show many structural similarities to humic acids but with a higher level of carboxyl and alcohol groups as noted above. Oxidative degradation studies, using CuO, show that lignin phenols are present in fulvic acids but these lignins are more highly altered compared to those of corresponding humic acids (Ertel and Hedges 1984). This relationship supports the degradative pathway in that fulvic acids are thought to be diagenetically downstream of humic acids.

In peats and poorly drained soils, fulvic acids contain significant levels of carbohydrate-like materials (Hatcher, Breger et al. 1983), no doubt derived from decomposing plant polysaccharides. It is not appropriate to consider carbohydrates in a chemical sense as humic substances, though they are co-isolated in the fulvic acid fraction and may even be linked to other "true" humic material as various chemical adsorbents have been used to remove carbohydrates from fulvic acid isolates (see Aiken, this volume). Thus, carbohydrates can contribute significantly to the apparent yield of fulvic acid isolates. It is important that the use of relative humic and fulvic acid yields for the inference of diagenetic trends (Stevenson 1982) be reevaluated with compositional data in mind. This is especially true for marine sediments where the data indicates a high carbohydrate content for fulvic acids.

Fulvic acids have been isolated, usually by use of macroreticular resins, from natural waters and seawater (Thurman 1985). The resins specifically exclude polysaccharide-like materials, so these fulvic acids are not contaminated with nonhumic substances. Though low levels of phenolic material can be detected in these aquatic fulvic acids, aliphatic materials are usually the dominant products. ^{13}C NMR data confirm this because aquatic fulvic acids show resonances predominantly in the paraffinic carbon region of the spectra (0–50 ppm). The NMR data show also an abundance of carboxyl carbon with little evidence for phenolic carbon except in aquatic fulvic acids from streams draining peat swamps (Thurman 1985). It is difficult to sort out the origin of aquatic fulvic acids because of the multitude of potential sources that exist in natural waters. The chemical composition of aquatic fulvic acid could thus be much more complex than that of fulvic acids in a soil or peat where sources of plant material are localized at the site of deposition.

Yields of humic isolates

While there are many factors that will influence the yield of a particular humic isolate (humin, humic acids, or fulvic acids) in sediments, one might expect that, in a degradative scheme, the yield of the various fractions might be related to the degree of degradation. Generally speaking such a trend is commonly observed, though many exceptions can be noted. In aerobic soils where degradation is greatest, the content of humic and fulvic acids is generally greater than that of the insoluble humin; in the most degraded aerobic systems, the fulvic acids predominate over humic acids (Stevenson 1982). The anaerobic environments of peats are generally known to be less degraded than aerobic soils, and the yields of soluble humic and fulvic acids are small compared to those of humins (Hatcher et al. 1985). Thus, a general positive trend exists between the yield of soluble humic substances and the degree of humification or degree of oxidative degradation.

Similar relationships exist among humic substances in marine sediments (Hatcher, Breger et al. 1983). Humin is generally the greatest component of marine sediments and humic acids are subordinate. In many instances fulvic acids are present in greater concentrations than humic acids. The ratio of fulvic acids to humic acids has been noted to decrease with increasing depth in sediments (Stevenson 1982), and this trend has been used to signify that fulvic acids were condensing to form humic acids. For reasons explained above, presence of nonhumic substances such as polysaccharides renders the above ratio meaningless unless atempts are made to segregate the carbo-hydrates from "true" humic matter. Nonetheless, it is likely that the absolute yield of "true" fulvic acids is significantly less than that obtained by simple extractions without fractionation. The fact remains, however, that the yield of humic acids in anaerobic marine sediments is less than yields in aerobic marine sediments. This is consistent with the expected trend if humic acids were being produced by oxidative degradation of humin, which is itself a modified biopolymer.

Electron spin resonance (ESR) studies

ESR has yielded some important information on the nature of free radicals in humic substances. Comparison of the ESR properties among humic fractions reveals that fulvic acids contain greater quantities of free radicals than humic acids (Schnitzer 1978). Furthermore, the free radical content was shown to be directly correlated to degree of humification or degradation, with the most humified soil samples having the greatest number of free radicals. These observations led Schnitzer (1978) to suggest that fulvic acids represent degraded humic acids and that the humification process is one of oxidative degradation. We are currently unaware of any reports of similar nature for marine humic substances.

Evolutionary trends in chemical composition

No doubt the best approach to the delineation of formation pathways for humic substances is to follow their chemical evolution over the course of degradation or humification. In some respects, we can call upon the natural system to provide information about humification occurring over a longer time span than could be approximated in the laboratory. Thus, studies of the geochemical evolution of both soil and sedimentary organic matter in depth profiles are particularly revealing of the natural humification process. Studies of this nature provided much of the fundamental evidence for Waksman's views on humification and have recently provided a great deal of evidence in support of the degradative concept of humification (Hatcher, Spiker et al. 1983; Orem and Hatcher 1987). It has been known for some time that microbiological degradation of woody plant tissue, under the moist aerobic conditions existing at the soil surface, involves extensive degradation of polysaccharides and lignin by numerous types of fungi and bacteria (Waksman 1938). If woody plant tissues are degraded anaerobically, as in a peat swamp or in a buried sediment, degradation proceeds at a slower pace. In such cases, carbohydrates are preferentially degraded and lignin-derived parts of the plant are selectively enriched (Hedges et al. 1985). It is this observation that led Fischer and Schrader (1921) to propose lignin as the "core" of the humic substances and coal.

Over the time scale of humification (years) the alteration of lignin can be minimal under anoxic conditions, but over geologic time (10^6 years) the lignin can be altered considerably by reactions that have been said to be analogous to those of the polycondensation models of humification (Flaig 1968). If this is, in fact, the case, then we might expect the lignin to be depolymerized to reactive phenolic intermediates that condense to yield humic acids that ultimately form progressively insoluble coal. Though it is possible that some components of coal may have gone through a soluble, small molecule intermediate, or gel phase, delicate cell structures in coal are often well preserved but nonetheless coalified (Stach 1975). In such cases it is attractive to assume that lignin was converted to coal without passing through a soluble phase. Recent studies of the coalification of xylem tissue using ^{13}C NMR and analytical pyrolysis (Hatcher et al. 1987) have confirmed that lignin can be altered by processes that are not unlike those proposed by Flaig (1968) for the alteration of lignin phenols. For example, transformation of lignin to lignite was shown to proceed through a catechol-like intermediate. In fact, catechol is a major product of analytical pyrolysis of lignitic xylem tissue (Hatcher et al. 1987). The significant aspect of this is that the transformation can occur with retention of cellular morphology and that lignin can be altered without being depolymerized. Therefore, if coalification (at low coal ranks where thermal degradation is minimal) can be viewed as humification then the above results would support the view that humification is a degradative process.

In soil and aerobic sediments, the alteration of lignin is much accelerated by oxidative reactions. The rapidity with which such alteration occurs may preclude identification of intermediates. However, in peat, where the level of degradation is low compared to aerobic soils, the degradation reactions could be sufficiently decelerated to detect humification trends as a function of depth. Orem and Hatcher (1987) examined humification in a peat from the Everglades, Florida with ^{13}C NMR. With increasing depth, the major biochemical process was degradation of carbohydrates as shown by Waksman (1938). Lignin-like materials appeared to be selectively preserved. Examination of the humin and humic acid isolates by ^{13}C NMR indicated that the lignin was the least altered in the humin fraction, but significant alteration of the lignin was observed in the humic acid fraction and even in dissolved organic matter (DOM) in pore fluids. This trend is suggestive of a mechanism in which lignin becomes modified by degradative processes in the peat to become incorporated into humic acid isolates and eventually into fulvic acids and DOM in pore fluids. Another interesting facet of the above studies was the presence of paraffinic carbons in humic, fulvic, and DOM fractions. The paraffinic components, believed to be from algal or microbial residues, were also selectively preserved with increasing depth in the peat.

Studies of humification of marine organic matter, or organic matter from nonvascular plants, have provided the bulk of information for proposed humification models in aquatic systems. However, the authors of these studies have generally explained their results using a polycondensation model (Stevenson 1982), though the data obtained could also be used in support of a degradative model.

Examination of the trends in the nature of insoluble humin in a marine algal sapropel from Mangrove Lake, Bermuda shows that selective preservation of a small but resistant component of the biomass can be used to explain humification in this system (Hatcher, Spiker et al. 1983). This resistant component, possibly a previously unknown biopolymer in algae or other microorganisms, is present in large amounts in the surface sediments and is selectively preserved with increasing depth to become the major fraction of the insoluble humin. This paraffinic-carbon-rich substance that contains about 4% nitrogen shows many spectral similarities of the paraffinic substance noted in the peat study described above. Thus, humification of nonvascular plant remains may follow the same process as that observed for vascular plant remains, with the early stages involving a selective preservation process and later stages involving modification of the biopolymer, possibly by oxidative processes (Vandenbrouke et al. 1985). Recent studies of the molecular weight relationships among humic substances in marine sediment cores (Cronin and Morris 1982) further support such a pathway for marine humic substances. This work suggests that an insoluble biopolymer is either formed or exists early in the depositional record.

Stable carbon isotopes

Carbon isotopes have played an important role in studies of the origin of humic substances. Organic matter in recent marine sediments, most of which is insoluble humin, is typically depleted in ^{13}C relative to local marine fauna and flora (Galimov 1980). In general, there is a systematic decrease of several per mil in the $^{13}C/^{12}C$ ratio along the series from fresh plants to humic substances. This decrease in ^{13}C could be due to the degradative loss of ^{13}C-enriched carbohydrates which account for a large fraction of the total carbon in fresh plant matter. Williams (1968) noted a several per mil ^{13}C depletion in marine DOM relative to marine plankton. He attributed this to the loss of ^{13}C-enriched carbohydrates as the plankton decompose. Spiker and Hatcher (1984) examined the quantitative relationship between $^{13}C/^{12}C$ and carbohydrate content in organic matter derived from algae. They concluded that the diagenetic loss of ^{13}C-enriched carbohydrates and selective preservation of ^{13}C-depleted organic matter during degradation resulted in a decrease of up to about 4 per mil in the $^{13}C/^{12}C$ ratio of the algal derived organic matter preserved (humin) in the sapropelic sediment. The magnitude of the decrease in the $^{13}C/^{12}C$ is a function of the relative ^{13}C enrichment of the carbohydrates.

Spiker and Hatcher (1987) found a similar relationship between $^{13}C/^{12}C$ and carbohydrate content in decomposed and low rank coalified woods. They concluded that as the carbohydrate content decreases the $^{13}C/^{12}C$ ratio in degraded wood decreases 1 to 2 per mil, approaching the value of the residual lignin. These results indicate that carbohydrate degradation products are lost and not incorporated into the aromatic structure as lignin is selectively preserved as humin during early diagenesis of wood.

PROBLEMS WITH UNDERSTANDING FORMATION MECHANISMS

As with many scientific hypotheses, the bulk of existing data supporting various proposed humification pathways can be used differently to support different theories. It must be emphasized that we do not currently have definitive, unambiguous evidence to solidly support any particular humification hypothesis.

Numerous factors are responsible for the lack of definitive evidence. First, there are severe problems with the operational definitions of humic substances that, at present, allow a variety of well-known biochemical compounds to be operationally defined as humic substances. Until this problem is resolved, it is doubtful that a unified theory of humification will emerge. Second, unless we can define, at both the bulk and molecular level, the chemical structural nature of humic substances, then it is virtually impossible to adequately define their origin. Recent studies have identified structural components (paraffinic components) that have not previously been known as components of humic substances. We must first reconcile whether these components are humic or nonhumic and, if they are humic, define their possible structural makeup. The third confounding factor is the lack of suitable natural systems for following the progress of humification. In this case, suitable would mean simple, where complicating factors such as variable plant sources and changing environmental conditions are not present. Of course this is where laboratory studies of humification are in order, but the time scales are usually prohibitive for such studies and we know little of how to replicate natural microbiological effects.

SUGGESTED FUTURE RESEARCH DIRECTIONS

It is clear that future work to resolve the question of how humic substances form must focus on eliminating the ambiguities of prior studies. We must first seek a clear operational definition of humic substances and then seek to define the chemical structural composition of the defined humic substances. Such studies should be made at the molecular level simultaneously using a variety of modern analytical capabilities, but we should be extremely cognizant of the bulk structure in relation to molecular level details. For example, we must be sensitive to what proportion of bulk material is represented by identified molecular level products obtained by various methods.

Inasmuch as laboratory humification experiments have provided valuable information, we should continue such studies keeping in mind the above needs to better characterize the end products. As stated by Flaig (1966), the experiments conducted to simulate humification allow no clear unambiguous determination of whether humic substances are formed by polymerization of small degraded molecules or by degradation of biopolymers. Of course, techniques available to us at present for characterization of products were not available to Flaig (1966); however, his experiments should be repeated with attention paid to providing unambiguous answers by use of modern methods.

Finally, it is important that we continue to search for ideal natural systems that allow us to follow the course of humification without complicating factors. Thus, peat deposits, soil horizons, or algal deposits that have been subjected to constant environmental conditions over extended periods of time might be ideally suited for such studies. We must seek to find such natural system where humification under one set of environmental conditions could be followed and modelled.

REFERENCES

Anderson, H.A., and Russell, J.D. 1976. Possible relationship between soil fulvic acid and polymaleic acid. *Nature* **260**: 597.

Crawford, R.L. 1981. Lignin Biodegradation and Transformation. New York: John Wiley.

Cronin, J.R., and Morris, R.J. 1982. The occurrence of high molecular weight humic material in recent organic-rich sediment from the Namibian Shelf. *Est. Coast. Shelf Sci.* **15**: 17–27.

Ertel, J.R., and Hedges, J.I. 1984. The lignin component of humic substances: distributions among soil and sedimentary humic, fulvic, and base-insoluble fractions. *Geochim. Cosmochim. Acta* **48**: 2065–2074.

Fischer, F., and Schrader, H. 1921. The origin and chemical structure of coal. *Brennstoff Chem.* **2**: 37–45.

Flaig, W. 1966. The chemistry of humic substances. In: The Use of Isotopes in Soil Organic Matter Studies, pp. 103–127. Oxford: Pergamon Press.

Flaig, W. 1968. Chemistry of humic substances in relation to coalification. In: Coal Science, pp. 58–68. Advances in Chemistry Series 55. Washington, D.C.: American Chemical Soceity.

Galimov, E.M. 1980. C^{13}/C^{12} in kerogen. In: Kerogen, Insoluble Organic Matter from Sedimentary Rocks, ed. B. Durand, pp. 271–299. Paris: Editions Technip.

Hatcher, P.G.; Breger, I.A.; Dennis, L.W.; and Maciel, G.E. 1983. Solid-state ^{13}C NMR of sedimentary humic substances: new revelations on their chemical composition. In: Aquatic and Terrestrial Humic Materials, eds. R.F. Christman and E.T. Gjessing, pp. 37–82. Ann Arbor: Ann Arbor Science Publishers.

Hatcher, P.G.; Breger, I.A.; Maciel, G.E.; and Szeverenyi, N.M. 1985. Geochemistry of humin. In: Humic Substances in Soil, Sediment, and Water, eds. G.R. Aiken, D.M. McKnight, R.L. Wershaw, and P. MacCarthy, pp. 275–302. New York: John Wiley & Sons.

Hatcher, P.G.; Lerch, H.E. III; Kotra, R.K.; and Verheyen, T.V. 1987. Pyrolysis/gas chromatography/mass spectrometry of a series of buried woods and coalified logs that increase in rank from peat to subbituminous coal. *Am. Chem. Soc. Div. Fuel Sci. Preprints* **32**: 85–93.

Hatcher, P.G.; Schnitzer, M.; Dennis, L.W.; and Maciel, G.E. 1981. Aromaticity of humic substances in soils. *Soil Sci. Soc. Am. J.* **45**: 1089–1094.

Hatcher, P.G.; Spiker, E.C.; Szeverenyi, N.M.; and Maciel, G.E., 1983. Selective preservation and origin of petroleum-forming aquatic kerogen. *Nature* **305**: 498–501.

Hedges, J.I.; Cowie, G.L.; Ertel, J.R.; Barbour, R.J.; and Hatcher, P.G. 1985. Degradation of carbohydrates and lignins in buried woods. *Geochim. Cosmochim. Acta* **49**: 701–711.

Largeau, C.; Casadevall, E.; Kadouri, A.; and Metgger, P. 1984. Formation of *Botryococcus*-derived kerogens. Comparative study of immature Torbanite and of the extant alga *Botryococcus braunii*. *Org. Geochem.* **6**: 327–332.

Martin, J.P., and Haider, K. 1971. Microbial activity in relation to soil humus formation. *Soil Sci.* **111**: 54–63.

Nip, M.; Tegelaar, E.W.; de Leeuw, J.W.; and Schenck, P.A. 1986. A new non-saponifiable highly aliphatic and resistant biopolymer in plant cuticles. *Naturwissenschaften* **73**: 579–585.

Nissenbaum, A., and Kaplan, I.R. 1972. Chemical and isotopic evidence for the *in situ* origin of marine humic substances. *Limnol. Oceanogr.* **17**: 570–582.

Orem, W.H., and Hatcher, P.G. 1987. Early diagenesis of organic matter in a sawgrass peat from the Everglades, Florida. *Int. J. Coal Geol.* **8**: 33–54.

Saiz-Jimenez, C.; Haider, K.; and Meuzelaar, H.L.C. 1979. Comparisons of soil organic matter and its fractions by pyrolysis mass-spectrometry. *Geoderma* **22**: 25–37.

Schnitzer, M. 1978. Some observations on the synthesis of humic substances. In: Proceedings of Research Conference 7, Laboratorio per la Chinnica del Terreno-Pisa, pp. 1–11. Pisa: Industria Tipografica C. Cursi & F.

Schnitzer, M., and Calderoni, G. 1985. Some chemical characteristics of paleosol humic acids. *Chem. Geol.* **53**: 175–184.

Schnitzer, M., and Khan, S.U. 1972. Humic Substances in the Environment. New York: Marcel Dekker.

Spiker, E.C., and Hatcher, P.G. 1984. Carbon isotope fractionation of sapropelic organic matter during early diagenesis. Org. Geochem. **5**: 283–290.

Spiker, E.C., and Hatcher, P.G. 1987. Carbon isotope fractionation in buried wood. *Geochim. Cosmochim. Acta* **51**: 1385–1391.

Stach, E. 1975. Coal Petrology. Berlin: Gebrüder Borntraeger.

Stevenson, F. 1982. Humus Chemistry. New York: John Wiley & Sons.

Thurman, E.M. 1985. Organic Geochemistry of Natural Waters. Boston: Niyhoff/ Junk Publishers.

Vandenbrouke, M.; Pelet, R.; and Debyser, Y. 1985. Geochemistry of humic substances in marine sediments. In: Humic Substances in Soil, Sediment, and Water, eds. G.R. Aiken, D.M. McKnight, R.L. Wershaw, and P. MacCarthy, pp. 249–274. New York: John Wiley & Sons.

Waksman, S.A. 1938. Humus, Origin, Chemical Compositions and Importance in Nature. Baltimore: Williams & Wilkins Co.

Williams, P.M. 1968. Stable carbon isotopes in the dissolved organic matter of the sea. *Nature* **219**: 152–153.

Humic Substances and Their Role in the Environment
eds. F.H. Frimmel and R.F. Christman, pp. 75–92
John Wiley & Sons Limited
© S. Bernhard, Dahlem Konferenzen, 1988.

Generation of Model Chemical Precursors

W. Flaig

*Formerly: Institut für Biochemie des Bodens,
FAL Braunschweig
Present address: Otto-Hahn-Strasse 132
8708 Gerbrunn, F.R. Germany*

Abstract. After a short discussion of the main steps of humification, the compounds formed by oxidative and reductive degradation of humic acids are described showing the substances which are useful for the synthesis of model humic acids.

The different reactions of natural-occurring lignin degradation products and microbial synthesized phenolics to polymers are mentioned. Possible reactions are investigated for model substances.

The formation of synthetic humic acids by oxidation of hydroxybenzenes in alkaline solution or by activity of peroxidase and H_2O_2 are explained by reactions of differently substituted polyphenols. A model for humic acids as redox systems is described.

An attempt to explain the function of nitrogen in synthetic humic acids formed by polymerization of model phenols in the presence of ammonia is made. The addition of amino acids or proteins to differently substituted quinones and their oxidative deamination is investigated as a function of the structure of the quinones. The mechanisms of these two reactions are elucidated by polarographic measurements.

INTRODUCTION

In order to demonstrate which model compounds can be used to elucidate the reactions in the course of humification, a summary scheme is given (Fig. 1), in which the transformation of organic constituents of plants and microorganisms is described.

According to our present knowledge the main plant consistuents, such as celluloses and hemicellulose (up to 40%), lignins (5–20%), and proteins (3–8%), are important initial materials for the processes of humification. In addition, several microbial synthesized phenols or amino sugar derivatives

75

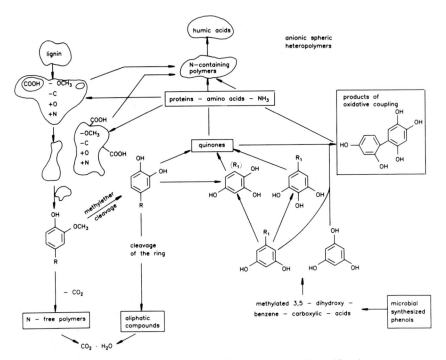

Fig. 1—Summary scheme of the processes of humification.

are also involved. Cellulose is mainly used as a source of energy by the microorganisms and is released largely as carbon dioxide in the atmosphere. The proteins decompose in the usual manner into amino acids, ammonia, and carbon dioxide.

The common concept for the synthesis of humic substances is reactions between phenolic compounds and proteins or their degradation products. There exist two sources of phenolic compounds: one produced by degradation of lignin, the other by microbial synthesis. Through the use of different analytical procedures several other compounds have been isolated from humic acids, such as carbohydrates, amino sugars, fatty acids, and others. In some cases they will be mentioned.

The high molecular weight lignins are degraded by microbial attack and/ or radical mechanisms to smaller and smaller molecular fragments which react with proteins and their degradation products to nitrogen-containing polymers, from which humic acids can be isolated in the usual way. During lignin degradation methoxyl groups and carbon content decrease, while oxygen and nitrogen content increase. Furthermore, the content of carboxy groups increases by cleavage of the rings and the C–C linkages between two monomers. Later this will be discussed in relation to model substances, which lignin degradation products with several monomer units can transform into substances with a quinonoid structure. After reaction with nitrogen compounds the newly formed substances could be considered as "precursors," or as humic acids "themselves," depending on the molecular weight.

Finally, low molecular weight decomposition products, such as phenolacrylic (R = – CH =CH – COOH) and phenolcarboxylic (R = –COOH) acids, can also be isolated in model experiments from rotted plant materials or directly from soils. It is known that they are not stable compounds, but they are continuously formed during humification (Maeder 1960). Details on these processes have been recently summarized by Haider (1987).

The partially methylated phenolic acids polymerize in the presence of oxidizing enzymes to form nitrogen-free polymers with a light yellow color. They are degraded by microorganisms serving as a carbon source.

A further important reaction is the cleavage of the methyl ether group, whereby phenols are formed with at least two hydroxy groups in orthoposition (o-position). They are very reactive and are formed via semiquinonoid radicals together with proteins and their degradation products, nitrogen-containing polymers, from which humic acids can be isolated.

In the case of o-dihydroxy compounds, oxidative ring cleavage occurs. The formed ketocarboxylic acids are degraded by microbial activity.

Another source of phenolic compounds comes from the synthesis by microorganisms, e.g., by *Epicoccum nigrum* (Haider and Martin 1967). This organism synthesizes 2-methyl-3,5-dihydroxy-benzene-carboxylic acid (R_1 = $-CH_3$) as well as 2-methyl-4,5-dihydroxy-benzene-carboxylic acid. By determining the transformation of these two acids, different phenolic compounds have been found with two or three hydroxy groups in the o-position. Only when these types of compounds are formed in the culture solution, and when quinonoid structures are possible, can humic acids such as melanins be synthesized. The orginal resorcinol derivatives and some of their transformation products of this type of compound do not participate alone in the synthesis of dark-colored products. They can, however, be included in the processes through oxidative coupling with compounds which can be transformed as such with quinonoid structure.

The transformation of 1,3-dihydroxy benzenes in 1,2,3- or mainly in 1,2,4-trihydroxy benzenes is not only important for the addition of amino acids but also (as mentioned below) for their oxidative deamination and the coupling of rings. The reactions mentioned occur only with quinonoid structures.

From the pool of nitrogen-containing polymers, formed from more or less different original phenolic and nitrogenous compounds, humic acids are extracted as anionic, spheric heteropolymers with weak alkaline solutions or strong polar solvents. One can imagine that to a certain extent the composition varies in the different monomers of lignins according to plant species, by the variable composition of the amino acids in the proteins, by the different microbial synthesized and participating compounds, as a result of more acidic or alkaline conditions, and by the varying properties of inorganic soil colloids or heavy metals, etc. Humic acids can be considered as a system composed of chemically related constituents which are formed by several principal reactions. In aquatic systems some other initial materials interact in the formation of humic substances.

DEGRADATION OF HUMIC ACIDS FOR INFORMATION ABOUT THE STRUCTURES OF CONSTITUENTS

Oxidative degradation

Schnitzer (1977) intensively investigated the oxidation of unmethylated and methylated humic acids from soils of different climatic regions. Apart from a larger number of aliphatic acids, Schnitzer found benzene- and phenol-carboxylic acids as well.

Spiteller (1981a) determined about 200–300 compounds after alkaline permanganate oxidation of methylated humic acids of a podsol profile with capillary gas chromatography mass spectrometry and found changes in the composition of the isolated compounds dependent upon the layer. The possibilities of participation of aromatic carboxylic compounds in the structure of humic acids were recently discussed by Hayes (1984).

The occurrence of most of the isolated polycarboxylic acids can only be explained through the assumption that with time polyaromatic compounds are formed. This may be deduced from the investigations of Ellwardt (1976), who determined different polycyclic aromatic hydrocarbons with an extraction procedure in deeper layers of peat (up to 2.5 m). Polycyclic aromatic hydrocarbons are therefore not contaminants of soil organic matter. The same type of compounds can also be found after zinc dust destillation of humic acids (cf. Schnitzer 1978; Hayes 1984).

Remark to the methylation of humic acids with diazomethane.   Spiteller (1981b) describes substituted pyrazols as constituents of humic acids' degradation products. After methylation of the humic acids, alkaline permanganate oxidation, and again methylation he isolated 1-methyl- and 2-methyl-pyrazole-dicarboxylic-(3,4)-dimethylester.

We found that differently substituted quinones add diazomethane besides the methylation of a hydroxy group (Flaig et al. 1963). In this way 4,7-diketo-tetrahydro-indazoles are formed (Fig. 2).

It can be concluded that vanillic units, in a more or less 1,4-quinonoid system of degraded lignin, add diazomethane even when it is connected with one or two carbon linkages to other units. The higher nitrogen content of

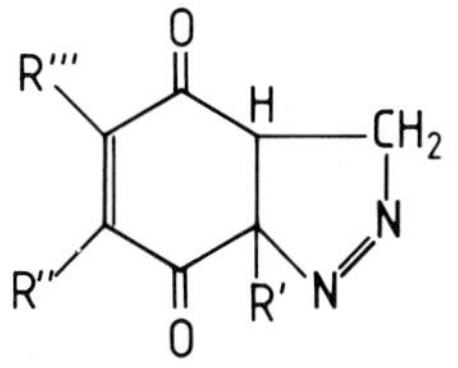

Fig. 2—4,7-diketo-tetrahydro-indazoles.

methylated humic acids may be explained by the addition of diazomethane as described. Syringic units cannot add diazomethane.

Reductive degradation of humic acid-like substances

Since Burges et al. (1964) introduced the use of sodium amalgam under nitrogen atmosphere for reductive degradation of humic substances, many investigations have been made in this field (e.g., Martin et al. 1974; Hayes 1984). Numerous phenolic compounds have been identified which cannot be transformed directly into quinones (OH-groups in 1,3- and 1,3,5-position). Furthermore, anthraquinones were also isolated from fungal polymers, which may be interesting for their physiological effect on plant growth (Flaig and Otto 1951).

Oxidative coupling. Among the phenolic compounds isolated by reductive degradation there are some which do not form quinonoid compounds under the oxidative conditions for the formation of humic acid-like polymers; their hydroxy groups are in the 1,3- or 1,3,5-position. In this case, reactions with other phenols occur only after oxidative hydroxylation or oxidative decarboxylation, as well as by oxidative coupling with phenols, which are able to form quinonoid systems. Some examples for oxidative coupling are given in Fig. 3.

3,5-Dimethyl-catechol reacts with 2,4-dimethyl-resorcinol to form 2,3,2',4'-tetrahydroxy-3,5,2',6'-tetramethyl-diphenyl (Baker and Miles 1955). 4-Methyl-catechol forms with phloroglucinol brown-colored polymers because in the case of unsubstituted compounds different possibilities of reactions exist. 4-Methyl-benzoquinone-1,2 reacts with 5-methoxy-6-methyl-resorcin to form a corresponding diphenyl derivative.

In the next example the influence of a substituent on the stability of the first reaction product—mostly a quinonoid compound—is demonstrated. The effect of the substitution with a methyl group is compared with that of a tertiary butyl group (tert.butyl). These experiments have been made because in oxidatively degraded lignin units a "substituent" may also be present, which causes sterical hindrance and whereby quinonoid structures may become more stable (cf. Fig. 7).

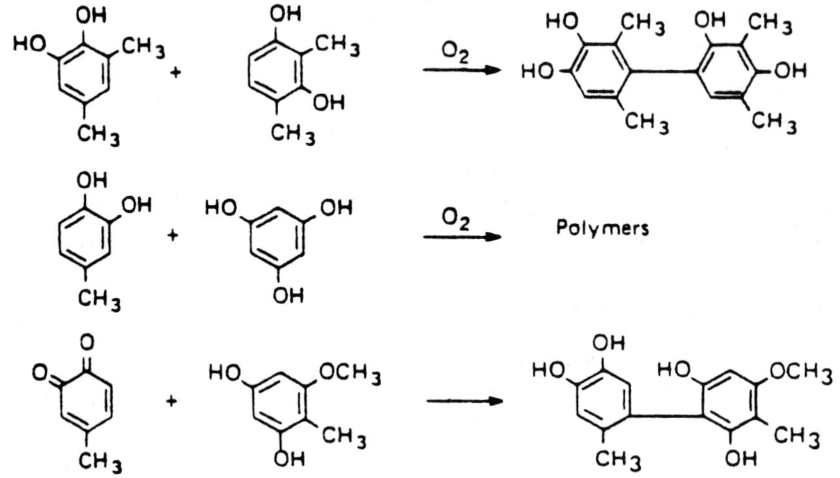

Fig. 3—Oxidative coupling of polyphenols.

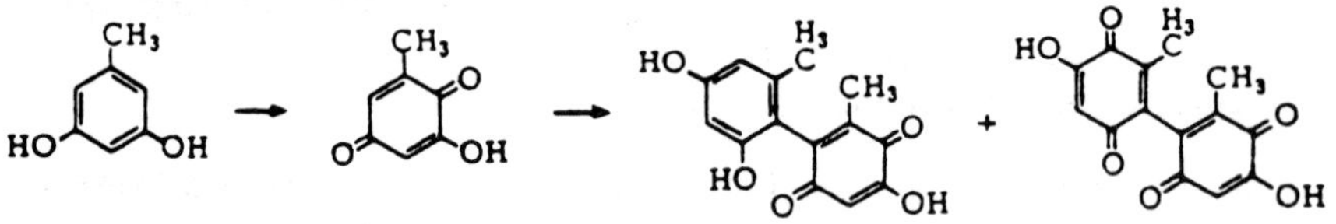

Fig. 4—Autooxidation of 5-methylresorcinol (orcinol).

The autooxidation of orcinol leads in aqueous solution through the intermediate 2-hydroxy-6-methyl-benzoquinone-1,4 to a mixture of two quinones (Fig. 4) that were isolated by Musso (1958). In contrast to this, in the case of 5-tert.butyl-resorcinol, the first reaction product 3-hydroxy-6-tert.butyl-benzoquinone can be isolated (Musso and Bormann 1965).

HOW FAR ARE CARBOHYDRATES ESSENTIAL CONSTITUENTS OF HUMIC ACIDS?

It is known that carbohydrates can be isolated from humic acids by acidic hydrolysis (Greenland and Oades 1975; Lowe 1978). There are two main possibilities for the presence of carbohydrates in the humic acids fractions. One is that they are coprecipitated, e.g., as polyuronic acids during isolation of humic acids. The other possibility is that they are included in the humic acids by acidic precipitation in the form of degraded lignin-carbohydrate complexes (Hayes et al. 1984).

With model studies it was found that alcoholic hydroxyl groups in carbohydrates may add to intermediates of quinone methide type to give p-hydroxy-benzyl-ether bonds (Taneda et al. 1985). It seems that the carbo-hydrates are bound in the humic acids by covalent bonds as proteins and amino acids.

THE NATURAL OCCURRENCE OF LOW MOLECULAR WEIGHT LIGNIN DEGRADATION PRODUCTS AND THEIR USE IN MODEL REACTIONS

Main steps in transformations

Most of the low molecular weight lignin degradation compounds, such as phenol acrylic or carboxylic acids mentioned in Fig. 1, can be isolated from soils, rotted plant materials, or peat. The main reactions of their transformation are:

1) shortening of the side chain of phenol acrylic acids and oxidation;
2) demethylation of the methoxy group;
3) hydroxylation and, in the case of 1,2-dihydroxy compounds cleavage of the ring, formation of keto acids; and
4) oxidative decarboxylation.

The last reaction is the start of the formation of humic acids by polymeriz-ation. In some cases dimerization products can also be isolated.

Dimerization reactions as models for the first steps of formation of humic substances.

The different substitution of vanillic acid with a methoxy group 3-position and protocatechuic acid with a hydroxy group has consequences for their dimerization. Two different diquinones are formed (Fig. 5). In the first case the connection of the rings occurs in the m-position of the methoxy group whereas in the second case it occurs in the p-position of the hydroxy group. This difference has an influence on the further polymerization. Finally, light-colored dehydrogenation polymers are formed from vanillic acid, which are not resistant against microbial attack. The polymerization of protocatechuic acid leads to dark-colored products, which are comparable to humic substances.

The probability that diphenylether takes part in the synthesis of humic acids is not large because they split off relatively easily in the presence of ferric ions. On the other hand, 2,3-dimethoxy-diphenylether-dicarboxylic acid-(5,4') and 2,3,2' trimethoxy-dephenylether-dicarboxylic acid-(5,4') were found in humic acids after methylation and alkaline permanganate oxidation with capillary gas chromatography (Spiteller 1981a).

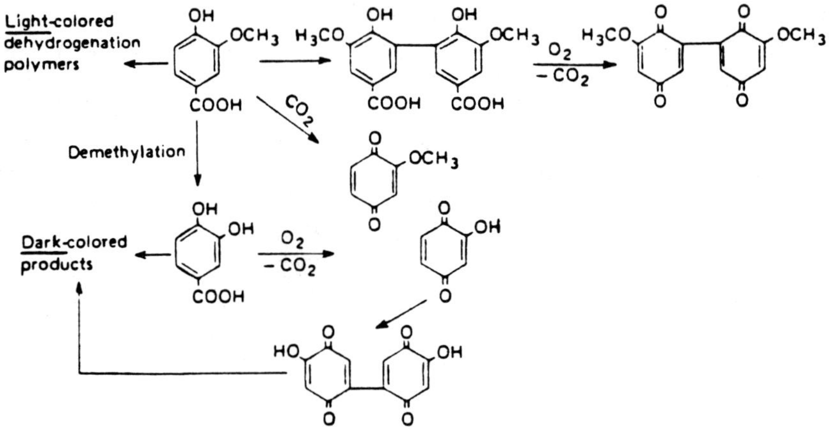

Fig. 5—Transformations of vanillic and protocatechuic acid in cultures of micro-organisms.

REMARKS ON SOME PROPERTIES OF PRECURSORS OF HUMIC ACIDS

Substitution and oxidation

The dependence of oxidation on substitution is briefly demonstrated. The substitution with ditert.butyl groups may be of interest as a model case for the course of the first steps of reactions during humification.

Through alkaline oxidation of hydroquinone or catechol, brown products are formed, but in the case of thymohydroquinone these are ether soluble. 3-Hydroxy-thymoquinone is oxidized to 3,6-dihydroxy-thymoquinone. 6-Hydroxy-thymoquinone is dimerized to the corresponding diquinone. Free rotation of the two quinone rings is impossible as was demonstrated by the spectrum in ultraviolet and visible light (Flaig and Salfeld 1958).

By substituting hydroquinone with tert.butyl groups in the 2,5-position, the oxidation in alkaline solution almost always leads quantitatively to 2,5-ditert.butyl-benzoquinone-1,4. Neither hydroxylation nor dimerization occurs (Flaig 1964).

o-,p-Tautomerism of 2-hydroxy-quinones-1,4 and 4-hydroxy-quinones-1,2

According to Fieser and Fieser (1954) one can suppose an equilibrium between 2-hydroxy-benzoquinone-1,4 and 4-hydroxy-benzoquinone-1,2, which plays a role during the synthesis of model humic acids as well as in nature. The equilibrium can be estimated from data of UV spectra (Ploetz 1954; Flaig et al. 1955b; Flaig and Salfeld 1958). One can also get further information from the IR spectra (Flaig and Salfeld 1959) and through measurements of the redox potentials (Flaig et al. 1968; Riemer 1970).

The second maximum of the two hydroxy-thymoquinones (OH group in 3- or 6-position) is shifted to longer wavelengths (about 400 μm), where normally the absorption of o-quinones occurs. The two hydroxyquinones are more effective in physiological experiments with plants than thymoquinone itself. 3,6-Dihydroxy-thymoquinone is ineffective. Some relationships exist between chemical constitution of quinones and their physiological effect (Flaig 1987).

Model for humic acids as a redox system

1,4-Diquinonyl-benzene was synthesized, whereby the two quinone rings are connected with a conjugated system (Ploetz 1955; Flaig 1955). The maxima in the UV spectra correspond to an o- and p-configuration. The tautomeric configurations are shown by the formula in Fig. 6. One form can be described as diradical, which is very reactive. Humic acids as "redox systems" are also discussed by Ziechmann (1977, 1981, and this volume).

Further transformations occur in the presence of water. Hydroxy-quinones are formed. Dehydrogenation leads to the corresponding hydroxy-quinones, which are acidic and could contribute to the acidic nature of humic acids. The mixture of the transformation products described in the formula yield spectra which correspond increasingly to those of natural humic acids and which are as a result uncharacteristic.

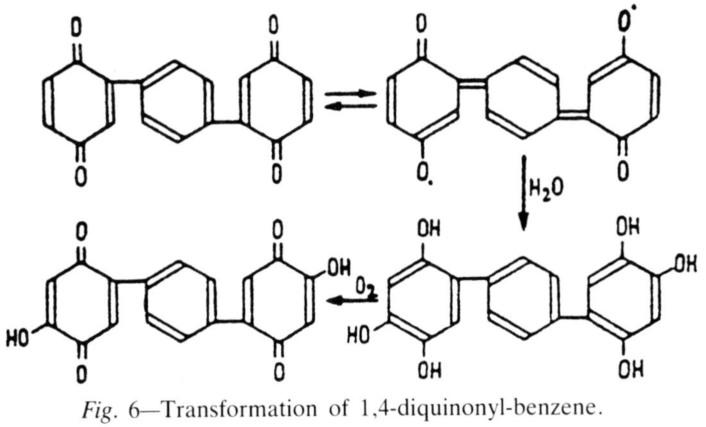

Fig. 6—Transformation of 1,4-diquinonyl-benzene.

Models for the cleavage of quinonoid rings

After finishing alkaline oxidation of hydroquinone, carbon dioxide evolves through acidification. The synthetic humic acids are separated. In the solution different degradation products of the initial compound, such as acetic and oxalic acid, can be identified. During formation of the synthetic humic acids a cleavage of the ring must have occurred.

Investigations with the model substance 2-hydroxy-3,3'-dinaphthyl-1,4,1',2'-diquinone show that the ring of o-quinonoid part, and not the p-quinonoid, is split off (Hooker and Fieser 1936). Different polyphenols have been substituted with tert.butyl groups to investigate the first steps of their oxidation. Above it was mentioned that 2,5-ditert.butyl-benzoquinone-1,4 is a very stable compound, unchangeable in alkaline oxidative solution. In contrast, the catechol and pyrogallol derivatives have shown a different behavior (Flaig et al. 1955a).

In the case of the catechol derivative, the deep red-colored 3,5-ditert.butyl-benzoquinone-1,2 was identified (Fig. 7). By further oxidation, through deep

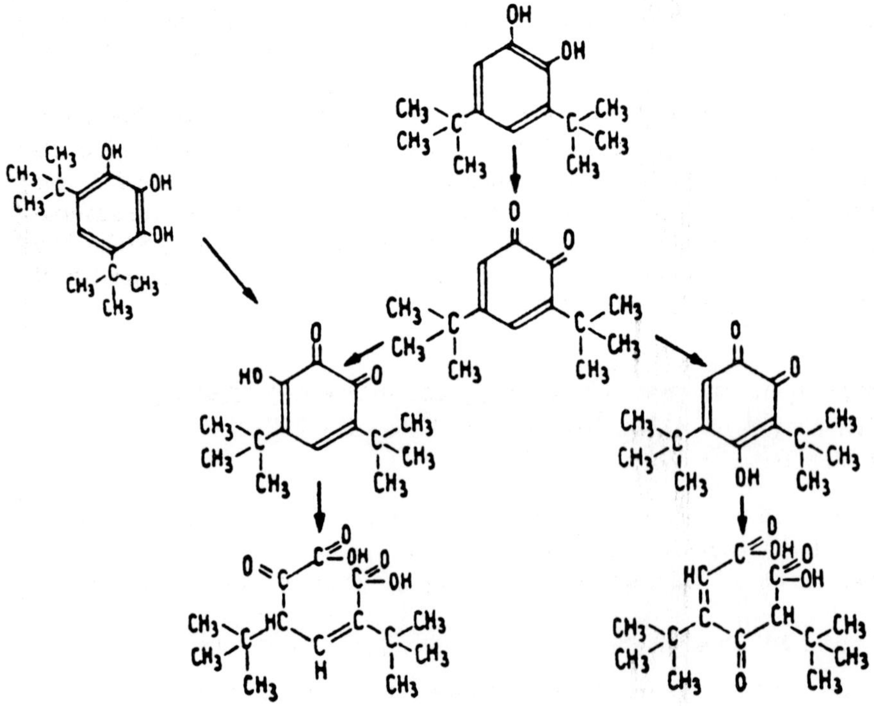

Fig. 7—Cleavage of the ring of catechol and pyrogallol substituted with two tert.butyl groups.

blue-colored intermediates, two crystallizable acids have been formed in nearly the same quantity, whereby one is 2,4-ditert.butyl-oxalocrotonic acid. According to the present knowledge it is probable that the other is 1,3-ditert.butyl-2-keto-buten-3-dicarboxylic-acid-1,4. In each case both dicarboxylic acids must have been formed by an isomeric hydroxy-hydroquinone (Flaig et al. 1954).

In the case of 4,6-ditert.butyl-pyrogallol the oxidative cleavage of the ring also occurs through the intermediate 3-hydroxy-4,6-ditert.butyl-benzoquinone-1,2. It seems to be that the cleavage of rings in initial materials occurs mainly between the two carbonyl groups of an o-quinonoid group in natural and synthetic precursors or in formed polymers. More details about constitution and transformation reactions of 3-hydroxy-benzoquinones-1,2 are described by Salfeld (1960) and Salfeld and Baume (1960, 1964).

Some special reactions of pyrogallol

In the case of catechol, hydroquinone, and hydroxy-hydroquinone the end product of oxidation under stronger conditions is always 2,5-dihydroxy-benzoquinone-1,4, which is a stable compound and more acidic than acetic acid. It liberates acetic acid from Ca-acetate. This result may be interesting for differentiation of phenolic and carboxy groups in humic acids (Schnitzer 1975).

In contrast to the above mentioned 1,2- and 1,4-dihydroxy-benzenes, pyrogallol and its derivatives form different types of dimerization products with its first oxidation product, such as 3-hydroxy-benzoquinone-1,2, and with differently substituted benzoquinones-1,2. The main results are summarized by Flaig (1969); special details are reported by Müller (1967). The variability of dimerization of pyrogallol may cause some special analytical data of pyrogallol synthetic humic acids (see below).

SYNTHETIC HUMIC ACIDS FROM HYDROXY-BENZENES BY OXIDATION IN ALKALINE SOLUTION OR BY MEANS OF PEROXIDASE + H_2O_2

Several authors (Eller 1921; Erdtman and Granath 1954; Flaig 1955) have studied the oxidation of hydroquinone in alkaline solution. The formed synthetic humic acids are more or less a polymer of hydroxy-benzoquinone-1,4 (Fig. 8).

According to several experimental data, it is supposed that

1) an o-p-tautomerism of the hydroxy-quinonoid structure exists;
2) a cleavage of the ring occurs to a smaller amount (the first step would be the formation of a dien structure which could react with the initial

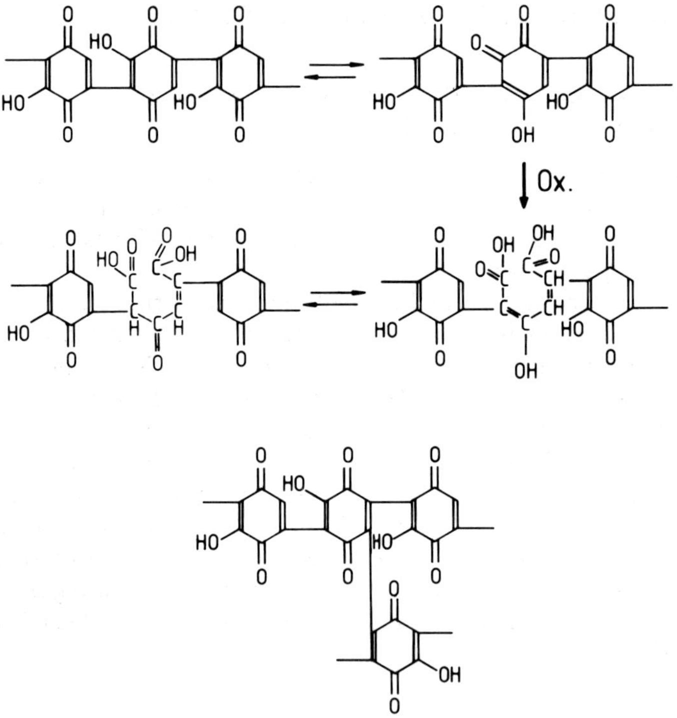

Fig. 8—Proposed structure of hydroquinone synthetic humic acids.

material oxidized to quinonoid structures);
3) the dicarboxylic acids are partially decarboxylated (therefore, CO_2 evolution occurs after acidification of the alkaline solution);
4) according to physical properties the polymer is branched;
5) inter- and intramolecular quinhydrones are formed (the polymer acts as electron exchanger and is a total redox system).

Schnitzer et al. (1984) investigated the formation of synthetic humic acids from catechol, hydroquinone, hydroxyhydroquinone, and pyrogallol through the use of horseradish peroxidase + H_2O_2. The synthetic humic acids were all relatively rich in free radicals. ^{13}C NMR spectra have indicated that they are compounds of a complex aromatic nature. Only the IR spectrum of the pyrogallol humic acid had a fine structure. This can be explained by the larger possibilities of dimerization reactions as mentioned above. The proposal for the structure assumes that the polymers are presumably also a source for the phenolic polycarboxylic acids formed by alkaline permanganate oxidation.

FUNCTION OF NITROGEN IN HUMIC SUBSTANCES

Nitrogen-containing synthetic humic acids by oxidation of hydroquinone in ammonia solution

After the uptake of up to 8 g atoms oxygen, the nitrogen-containing synthetic humic acids can be isolated with yields of about 90% (in sodium alkaline solution about 10–20% according to the reaction conditions). After acidification no CO_2 is evolved.

When formaldehyde is simultaneously present in the ammonia solution the same nitrogenous humic acids are formed, while in the case of sodium alkaline solution a resin of the type phenol-formaldehyde is found. Ammonia seems to block the reactive centers more than formaldehyde does.

The nitrogen content of humic acids has an influence on the internal structure of the molecule. With increasing nitrogen content the diameters of the intramolecular voids of the structure decrease; this can be concluded by determining the specific viscosity (Flaig et al. 1975). The results are in agreement with viscosimetric measurements of Orlow and Gorskova (1965). They found that the humic acids from chernozem with a higher nitrogen content were more compact than those of a sod-podsol soil (Schnitzer 1978; Orlow 1985).

Reactions between amino acids or proteins with differently substituted quinones as model precursors of naturally occurring phenolic compounds in soil

The substituents in the molecules of the quinonoid oxidation products formed during humification change the distribution of the electron density at the carbon atoms of the ring and therefore their chemical reactivity. So, it is possible that a substitution with the same substituents, but in other positions in the ring, can affect different reactions with amino acids. The electron density can be calculated from physical data. The investigated quinones can be divided into groups, each yielding different reactions—as described—or having physiological effects on plant growth (Flaig 1987).

A large number of publications describe the reactions of quinonones with amines or amino acid esters as model substances. There exists a principal difference in the chemical behavior of amino acid esters and the amino acids themselves. Amino acid esters react like aliphatic or aromatic amines. In the case of amino acids addition products are formed, with some quinones independent of their substitution, which are in a quinol-quinone equilibrium. The OH group of the COOH group interacts with a carbonyl group of the quinone (Riemer 1970).

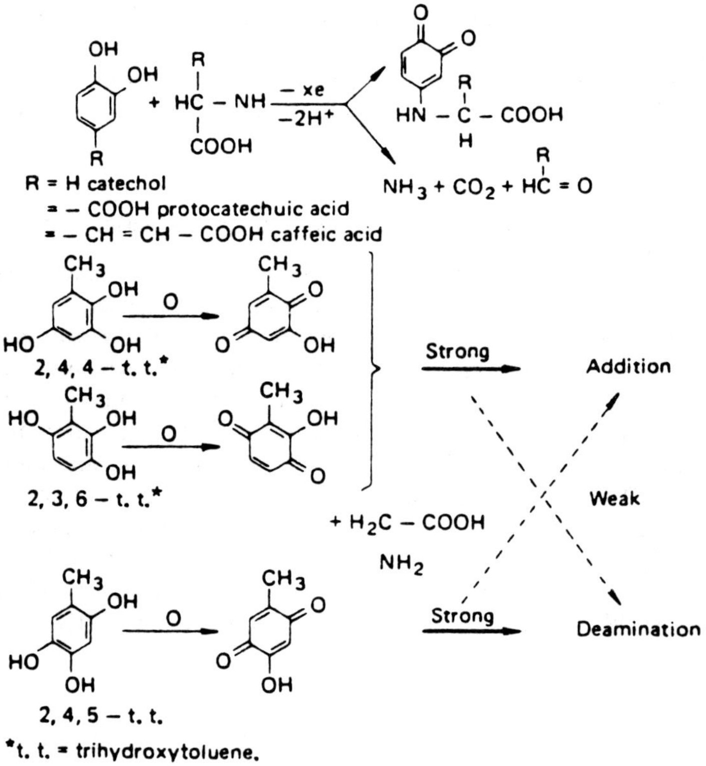

Fig. 9—Addition and/or oxidative deamination of amino acids by phenolic or quinonoid compounds in the presence of phenoloxidases.

Besides the formation of addition products a second reaction, the deamination of the amino acids, was observed. In this case the amino acid is decomposed in CO_2, NH_3, and an aldehyde compound (Fig. 9).

With differently ^{14}C-labelled amino acids, special processes could be followed (Haider et al. 1965).

Lignin degradation products such as vanillic acid or syringic acid, which are not yet transformed in quinonoid systems, do not add or deaminate amino acids. Only after transformation (e.g., vanillic acid in methoxy-benzoquinone-1,4) do addition reactions occur.

In some cases (e.g., 2-hydroxy-3-methyl-benzoquinone-1,4 and 2-hydroxy-6-methyl-benzoquinone-1,4) the addition is stronger than the deamination and vice versa (e.g., 2-hydroxy-5-methyl-benzoquinone-1,4) (Flaig and Riemer 1971).

Peptides and proteins are also added. Through hydrolysis of the addition products with 6 mol/l HCl all the amino acids of the peptides or proteins could be determined again, with the exception of the N-terminal one which is connected with the ring.

The course of the reactions, which occur during oxidative addition or deamination, were followed by polarographic methods in a nonenzymatic system.

The reaction mechanism of oxidative deamination of amino acids

According to extensive polarographic measurements the mechanism of the nonenzymatic oxidative deamination of amino acids, which is catalyzed by quinones, is described as follows. The addition of the amino acid to the quinone leads to a hydro-hydroxyquinone addition product, which is oxidized by addition of oxygen to a system with a stronger o-quinonoid character. This fact can be concluded by the positive shift of the potential (from -280 mV to -166 mV). Two further steps in the rearrangement of the double bond could be supported by polarographic measurements. Through further oxidation the hydroquinone compound is oxidized to the corresponding quinone, then to glyoxylic acid, ammonia, and finally to the initial compound, 2-hydroxy-5-methyl-benzoquinone-1,4. The latter adds again glycine and the reaction cycle goes on (Flaig and Riemer 1971).

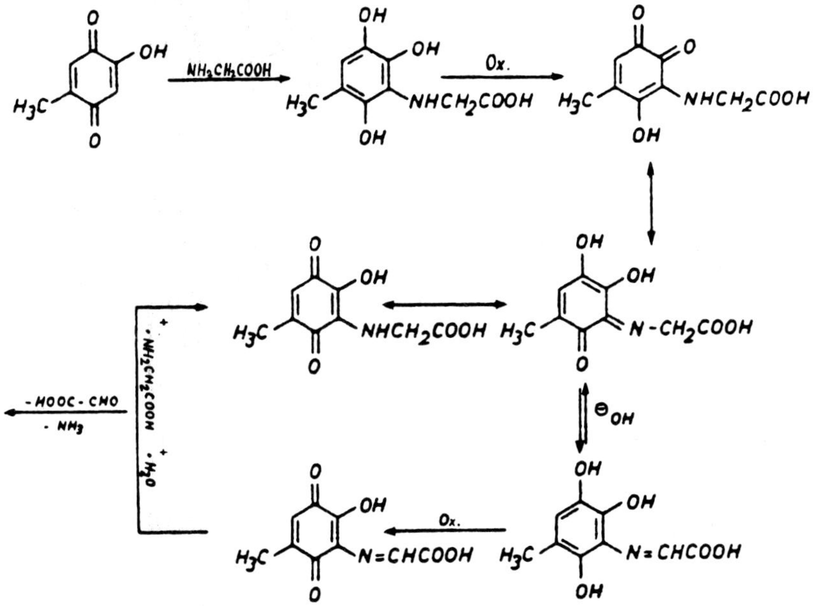

Fig. 10—The reaction mechanism of the oxidative deamination of glycine by 2- hydroxy-5-methyl-benzoquinone-1,4.

The reaction scheme of Trautner and Roberts (1950) is often used for the description of the mechanism of the deamination of amino acids. In this scheme the intermediate formation of a quinone-immine and an addition of a second amino acid is included. These two reactions could not be confirmed by polarographic measurements. Therefore, the oxidative deamination cannot be explained by the scheme proposed by Trautner and Roberts.

REFERENCES

Baker, W., and Miles, D. 1955. Red color given by coal-tar-phenols and aqueous alkalies - isolation of a quinone derived from a tetrahydroxytetramethyl biphenyl. *J. Chem. Soc.* **1955**: 2089–2096.

Burges, A.; Hurst, H.M.; and Walkden, B. 1964. The phenolic constituents of humic acid and the relation to the lignin of the plant cover. *Geochim. Cosmochim. Acta* **28**: 1547–1554.

Eller, W. 1921. Künstliche und natürliche Huminsäuren. *Brennstoffchem.* **2**: 129–133.

Ellwardt, P.-C. 1976. Zum Nachweis von polycyclischen Kohlenwasserstoffen mit und ohne cancerogene Wirkung in Torfen im Vergleich zum Vorkommen in Böden und Komposten. *Telma* **6**: 135–144.

Erdtman, H.G.H., and Granath, M. 1954. Studies on humic acids. 5. The reaction of p-benzoquinone with alkali. *Acta Chem. Scand.* **8**: 811–816.

Fieser, L.F., and Fieser, M. 1954. Lehrbuch der organischen Chemie, p. 817. Weinheim: Verlag Chemie.

Flaig, W. 1955. Beiträge zur Chemie der Modellsubstanzen von Huminsäuren. *Landw. Forschung* **6**: 94–101.

Flaig, W. 1964. Chemische Untersuchungen an Huminstoffen. *Zeitschr. Chem.* **4**: 253–265.

Flaig, W. 1969. Fortschritte der Huminstoff-Chemie. *Landbauforschung Völkenrode* **19**: 53–66.

Flaig, W. 1987. Effects of some organic substances on plant growth and metabolism. *Int. Humic Subst. Soc.* **3**, in press.

Flaig, W.; Beutelspacher, H.; Riemer, H.; and Kaelke, E. 1968. Einfluss von Substituenten auf das Redoxpotential substituierter Benzochinone-1,4. *Lieb. Ann. Chem.* **719**: 96–111.

Flaig, W.; Beutelspacher, H.; and Rietz, E. 1975. Chemical composition and physical properties of humic substances. In: Soil Components, ed. J.E. Gieseking, pp. 168–174. Berlin: Springer-Verlag.

Flaig, W., and Otto, H. 1951. Zur Kenntnis der Huminsäuren. III. Untersuchungen über die Einwirkung einiger Chinone als Modellsubstanzen der Auf- und Abbauprodukte von Huminsäuren sowie einiger Redoxsubstanzen auf das Wachstum von Pflanzenwurzeln. *Landw. Forschung* **3**: 66–89.

Flaig, W.; Ploetz, T.; and Biergans, H. 1955a. Zur Kenntnis der Huminsäuren. XIV. Bildung und Reaktionen einiger Hydroxychinone. *Lieb. Ann. Chem.* **597**: 196–213.

Flaig, W.; Ploetz, T.; and Küllmer, A. 1955b. Über einige Ultraviolettspektren einiger Benzochinone. *Zeitschr. Naturfor.* **108**: 668–676.

Flaig, W., and Riemer, H. 1971. Polarographische Untersuchungen zum Verhalten von Trihydroxytoluolen bei der Reaktion mit Glycin unter oxydierenden Bedingungen. *Lieb. Ann. Chem.* **746**: 81–85.

Flaig, W., and Salfeld, J.-C. 1958. UV-Spektren und Konstitution von p-Benzochinonen. *Lieb. Ann. Chem.* **618**: 117–139.

Flaig, W., and Salfeld, J.-C. 1959. Infrarotspektren von p-Benzochinonen. *Lieb. Ann. Chem.* **626**: 215–224.

Flaig, W.; Salfeld, J.-C.; and Rabien, G. 1963. Die Bildung von 1,2-Di-(benzochinon-2',5')-äthanen aus Additionsprodukten von Diazomethan an p-Benzochinone. *Makro. Chem.* **69**: 206–212.

Flaig, W.; Schulze, H.; Küster, E.; and Biergans, H. 1954. Zur Chemie der Huminsäuren. *Landbouw. Tijdschr.* **66**: 392–407.

Greenland, D.J., and Oades, J.M. 1975. Saccharides. In: Soil components, ed. J.E. Gieseking, pp. 213–262. New York, Heidelberg, Berlin: Springer-Verlag.

Haider, K. 1987. Synthesis and degradation of humic substances in soil. *Int'l. Soc. Soil Sci. Trans.*, XIII Congress, vol. 6, pp. 644–656.

Haider, K.; Frederick, L.R.; and Flaig, W. 1965. Reactions between amino acid compounds and phenols during oxidation. *Plant Soil* **22**: 49–64.

Haider, K., and Martin, J.P. 1967. Synthesis and transformation of phenolic compounds by *Epicoccum nigrum* in relation to humic acid formation. *Soil Sci. Am. Proc. 31*, **6**: 766–772.

Hayes, M.H.B. 1984. Structures of humic substances. In: Organic Matter and Rice, Internat. Rice Res. Inst. pp. 93–116. Philippines: Los Baños Laguno.

Hayes, M.H.B.; Dawson, J.E.; Mortensen, J.L.; and Clapp, C.E. 1984. Electrophoretic characteristics of extracts from sapric histosol soils. Int. Humic Substances Society, 2nd International Conference, M.H.B. Hayes and R.S. Swift, eds., pp. 31–41. Birmingham, U.K.

Hooker, S.C., and Fieser, L.F. 1936. Concerning Wichelhaus' "Di-ß-naphthoquinone Oxide." *J. Am. Chem. Soc.* **58**: 1216–1223.

Lowe, L.E. 1978. Carbohydrates in soil. In: Soil Organic Matter, eds. M. Schnitzer and S.U. Kahn, pp. 65-93. Amsterdam, Oxford, New York: Elsevier.

Maeder, H. 1960. Chemische und pflanzenphysiologische Untersuchungen mit Rottestroh. Dissertation, Univ. Giessen, FRG.

Martin, J.P.; Haider, K.; and Saiz-Jimenez, C. 1974. Sodiumamalgam reductive degradation of fungal phenolic polymers, soil humic acids, and simple phenolic compounds. *Soil Sci. Soc. Am. Proc.* **38**: 360–365.

Müller, H.H. 1967. Additionsverbindungen aus o-hydroxy-o-benzochinonen und alkyl-substituierten o-Benzochinonen sowie deren Umlagerungsprodukten. Dissertation, Techn. Univ. Braunschweig, FRG.

Musso, H. 1958. Über Orceinfarbstoffe. 7. Synthese, Konstitution und Lichtabsorption des Henrichschen Chinons. *Chem. Ber.* **91**: 349–363.

Musso, H., and Bormann, D. 1965. Über Orceinfarbstoffe XXII. Die Autoxydation des 5-tert.Butyl-resorcins. *Chem. Ber.* **98**: 2774–2796.

Orlow, D.S. 1985. Humic acids of soils. Russian Translation series 35a, p. 378. Rotterdam: A.A. Bakkma.

Orlow, D.S., and Gorskova, E.I. 1965. Size and shape of particles of humic acids from chernozem and sod-podzolic soils (Russ.). Nauchn. Dokl. Vysshei Shkoly, Bol. Nauki, No. 1.

Ploetz, T. 1954. Zur Bestimmung des Redoxpotentials reversibler organischer Systeme. *Z. Naturforsch. Bd.* **9b**: 753–755.

Ploetz, T. 1955. Polymere Chinone als Huminsäuremodelle. *Z. Pflanz. Boden.* **69**: 50–58.

Riemer, H. 1970. Polarographische Untersuchungen über Beziehungen zwischen Struktur und Reaktivität bei Benzochinonen. Dissertation, Techn. Univ. Braunschweig, FRG.

Salfeld, J.-C. 1960. Über die Oxydation von Pyrogallol und Pyrogallolderivaten. II. Die Konstitution dimerer 3-Hydroxy-o-benzochinone. *Chem. Ber.* **93**: 737–745.

Salfeld, J.-C., and Baume, E. 1960. Über die Oxydation von Pyrogallol und Pyrogallolderivaten. III. Die Umlagerung von dimerem 3-Hydroxy-4,6-di-tert.butyl-o-benzochinon durch Alkali. *Chem. Ber.* **93**: 745–751.

Salfeld, J.-C., and Baume, E. 1964. Über die Oxydation von Pyrogallol und Pyrogallolderivaten. IV. Die Konstitution der Purpurogallin-carbonsäure-(9). *Chem. Ber.* **97**: 307–311.

Schnitzer, M. 1975. Chemical, spectroscopic, and thermal methods for the classification and characterization of humic substances. In: Humic Substances, Their Structure and Function in Biosphere, eds. D. Povoledo and H.L. Golterman, pp. 308–309. Wageningen: Centre for Agric. Publ. and Docum.

Schnitzer, M. 1977. Recent findings on the characterization of humic substances extracted from soils from widely differing climatic zones. Soil Organic Matter Studies, IAEA, Vienna **2**: 117–132.

Schnitzer, M. 1978. Humic substances: chemistry and reactions. In: Soil Organic Matter, eds. M. Schnitzer and S.U. Khan, pp. 21–22, 34–35. New York: Elsevier.

Schnitzer, M.; Barr, M.; and Hartenstein, R. 1984. Kinetics and characteristics of humic acids produced from simple phenols. *Soil Biol. Biochem.* **16**: 371–375.

Spiteller, M. 1981a. Kapillar-GC-MS von Huminsäureabbauprodukten eines Podsols. *Z. Pflanz. Boden.* **144**: 472–485.

Spiteller, M. 1981b. Substituierte Pyrazole als Bestandteile von Huminsäureabbauprodukten. *Z. Pflanz. Boden.* **144**: 500–504.

Taneda, H.; Hosoya, S.; Nakona, J.; and Chang, H.-M. 1985. Behaviour of lignin-hemicellulose linkages in chemical pulping. Int. Symp. on Wood and Pulping Chem., Vancouver, B.C., August 26–30, 1985, Poster Presentations, pp. 117–118.

Trautner, E.M., and Roberts, E.A.H. 1950. The chemical mechanism of the oxidative desamination of amino acids by catechol and polyphenoloxidase. *Aust. J. Sci. Res. Ser. B* **3**: 356–380.

Ziechmann, W. 1977. Molekülkomplexe bei Huminstoffen durch ϵ-Donator und Akzeptor-Strukturen. *Z. Pflanz. Boden.* **140**: 133–150.

Ziechmann, W. 1981. Über Eigenschaften und Aufbau der Huminstoffe. *Telma* **11**: 159–176.

Humic Substances and Their Role in the Environment
eds. F.H. Frimmel and R.F. Christman, pp. 93–103
John Wiley & Sons Limited
© S. Bernhard, Dahlem Konferenzen, 1988.

Generation in Controlled Model Ecosystems

F.K. Pfaender

Department of Environmental Sciences and Engineering
University of North Carolina
Chapel Hill, NC 27514, U.S.A.

Abstract.   This paper reviews the need for new systems to produce humic compounds for use in a number of different types of investigations. The goal is to include more of the natural ecosystem reactions into the synthesis. Three approaches are suggested. These include simple batch reactors, soil columns, and chemostats. The possible options for using each are discussed.

INTRODUCTION

Over the last fifty years many workers have investigated various aspects of humic acid structure and formation using model humic-like compounds. A wide variety of percursor compounds have been used, ranging from individual compounds to complex mixtures of phenolics. The assumption used in many of these studies is that humics form from the polymerization of monomeric units. The results of these studies are discussed by Flaig (this volume) and have been reviewed by Flaig et al. (1975) and Stott and Martin (1987). Since natural humic acids may be formed from a variety of natural materials and can form in many different environments under many different kinds of conditions, there is great variability in the composition of natural humics. Models permit processes to be studied with better defined materials than natural humic acids. Models are also useful for studies of pesticide binding to soil constituents, of the immobilization and stabilization of enzymes, and of other materials added to the soil, such as crop residues. A more recent concern has been the nature of the reactions of humic material with a variety of disinfectant chemicals, which leads to the production of compounds of

93

great health and environmental significance. The use of model humic acids has lead to a greater understanding of the possible processes involved in humic acid formation, the potential starting materials involved, and the incorporation of soil organic material into humic polymers.

The problem in studying model humics is the same encountered in the study of natural humics. The polymers produced are complex and contain many different individual components. Sorting out what those components are and their significance to the overall structure remains a major problem in understanding both natural humics and the extent to which any model or particular formation mechanism mimics what happens in nature. One approach in generating synthetic humic acids that has received little attention to date is the use of a model ecosystem that can be controlled in several ways and allows humics to form under conditions that can be made much closer to those that exist in nature. Since it is clear that biological processes are important in the formation of humic acids in nature, the use of model ecosystems allows the biological component to be incorporated into the formation of the model compounds. This can be done without making basic assumptions about particular formation mechanisms.

Model ecosystem is clearly a term that can mean many things to many different people. In its simplest form it can be a flask or beaker into which some sample of the environment is placed, the sample is then manipulated to achieve some desired outcome. In many ecological or environmental studies much more complex model ecosystems have been used to simulate both the fate and transformation of a wide variety of compounds. The fungal culture systems used by Saiz-Jimenez et al. (1975) and Filip et al. (1974) are examples of simple model ecosystems that have been used to study the formation of humic acid-like pigments by certain fungi. There are some distinct advantages, as well as disadvantages, to the use of controlled model ecosystems for the production of model humic acid. In a model ecosystem we are attempting to simulate nature and the reactions that would proceed under natural conditions. Therefore, the humic acid that would be formed should be much more like a real humic acid than those formed as the result of chemical reactions. The model ecosystem can be constructed to be as simple or complex as the researcher might want. With increased complexity would come the ability to manipulate the system during the formation of the humic acid which would allow various biodegradation reactions or precursor molecules to be added, as well as different biological and abiotic catalysts. The system can also be controlled in terms of temperature, pH, protein or nitrogen content, or any variable deemed important.

A disadvantage of model ecosystems is the fact that the product formed may be considerably more complex than would be formed as the result of chemical synthesis methods. This may make isolation and analysis somewhat more difficult. It is also quite likely that under conditions much closer to

Table 1 Humic acid precursors.

Precursor	Source	Process
Phenolics	lignin sugars & polysaccharides	microbial degradation microbial synthesis
Polysaccharides	plant materials soil	microbial degradation microbial synthesis
Amino sugars	microbial cells	microbial degradation
Amino acids	plant & microbial proteins	microbial degradation

nature the rate of product formation would be slower than in traditional chemical synthesis techniques. It is also likely that using more or less natural conditions to produce the humic compounds may be somewhat less reproducible than humics from models produced by chemical synthesis.

This paper is not intended as a comprehensive review of previous efforts to produce model humic acids. That topic is addressed by others in this volume. After a brief discussion of the various components that should be considered for inclusion in any model ecosystem, this chapter will focus on some approaches that may be usefully applied to the generation of humic acids in controlled model ecosystems. I also make no presumptions as to the actual mechanism in nature. Model ecosystems may be a good research approach to distinguish between the suggested major mechanisms.

MODEL HUMIC ACID PRECURSORS

The formation of humic acids in nature is a complex and dynamic process that is likely to involve precursor molecules from many different sources, as shown in Table 1. Clearly, lignins can serve as a major source of phenolic compounds that eventually polymerize to form humic polymers (Flaig et al. 1975; Haider et al. 1975; Ladd and Butler 1975), or as a starting material that is degraded over time to produce various humic materials. As a result of the activity of soil microorganisms phenolic substances can be released during the degradation of lignins (Flaig et al. 1975). These phenolics could be further transformed by soil microorganisms into somewhat simpler phenolics which undergo microbial oxidation reactions and become linked into humic acid molecules. Alternatively, the biopolymers may be slowly altered by biotic and abiotic processes that result in changes to the structure. It is also apparent that lignin-derived molecules, at several different stages of decomposition, can undergo linkage into the humic polymer (Ladd and Butler 1975; Martin and Haider 1980b). It now seems clear that phenolic compounds in soil arise from sources other than lignin degradation. Soil fungi produce melanins, which are secondary metabolites formed during the metabolism of sugars and polysaccharides, that are complex phenolic polymers. Recent studies indicate that as many as 40 different phenolic and aromatic compounds can be produced by fungi and that most of these can be found as constituents of the melanins (Haider et al. 1965; Saiz-Jimenez et al. 1975). As with the lignin-derived polymers, these melanins may undergo microbial transformation in the soil before incorporation into soil humics.

The polysaccharide fraction of humics most likely originates either through microbial synthesis (Cheshire 1977) or from plant polysaccharides altered by microorganisms. Soil polysaccharides also may contain amino acids and amino sugars. The amino sugars almost certainly come from microbial cell components, one of the major sources in soil. It has been shown that amino sugar and amino acid units may be incorporated into humic acid-type polymers (Bondietti et al. 1972; Martin et al. 1978). It has also been shown that amino acids can react with phenols to form covalent linkages that incorporate the nitrogen into humic acid molecules (Haider et al. 1965; Ladd and Butler 1966). Organic phosporus can also be incorporated into humic acids as a result of covalent linkages formed during the polymerization process (Brannon and Sommers 1985).

With such a wide variety of precursor molecules available for the polymerization reactions that have been used to form model soil humics, it is no wonder that the polymers are as complex as has been shown. Given the fact that the nature and concentration of the various precursors can vary from environment to environment, largely due to the nature and concentration of the source material available and the activity of the soil microbes, it is clear why humic acids can vary as much as they do from site to site. From the standpoint of constructing a controlled model exosystem it is also clear that there are many potential sources and types of compounds to be considered for inclusion in the model ecosystem.

MODEL HUMIC ACID RESEARCH

For several decades researchers have been producing model humic acids by a variety of different techniques. These techniques all involve some type of oxidative polymerization, usually under alkaline conditions. All of these polymerization reactions appear to result in the formation of compounds that, to at least some extent, are comparable to natural humic acids. Most comparisons have been based upon such parameters as elemental composition, spectral measurements, functional group analysis, and, more recently, ^{13}C NMR and GC-MS analysis. While all the polymers produced in these reactions are nominally similar, it is clear that certain combinations of precursors and catalysts produce compounds more similar to natural humic

Table 2 Summary of model humic acid catalysts.

Oxidative Catalyst	Precursors	Reference
AgO	catechol, benzoquinone	Mathur and Schnitzer 1978
	catechol, protocatechuic acid	Goh and Stevenson 1971
$K_2S_2O_8$	various phenolics	Mathur and Schnitzer 1978
Phenoloxidase	phenols, amino acids	Haider et al. 1965
Mushroom phenolase	phenols, amino acids	Martin et al. 1972
Peroxidase-H_2O_2	phenols	Martin and Haider 1980a
Autooxidation	phenols, amino acids	Martin et al. 1972
Ca-illite	phenols, amino acids	Wang et al. 1985
Metal oxides	hydroquinone	Shindo and Huang 1985
H_2O_2-$FeSO_4$	phenols	Liogon'kii et al. 1985
Electrolysis	lignite	Lalvani et al. 1986

acids (Mathur and Schnitzer 1978; Martin and Haider 1980a). The actual degree of similarity in molecular structure between the synthetically produced and natural humic acids is, of course, unknown since there is no reasonable way, at present, to determine the overall molecular structure of either. In any case, it is likely to be highly variable from environment to environment and time to time. This raises the question as to how good a model the model humic acids may be, especially since they generally are based on the assumption that humics form as a result of polymerization of monomers rather than through degradative processes. There are certain factors that need to be considered if we are to use the research that has been conducted on model humic acids as a basis for formulating controlled model ecosystems.

In most studies of the formation of model humic acids single oxidative catalysts are used (Table 2). In nature it is likely that there are several catalysts present simultaneously including various enzymes, such as phenol oxidases and peroxidase, as well as several nonenzymatic catalysts, such as metal oxides and clays. It is therefore likely that in nature the reactions will be much more complex and lead to a significantly more diverse polymer than might be formed in model systems. The concentrations of reactants in nature are probably considerably lower than used in studies of the formation of model polymers. In addition to producing lower rates of formation, the concentration effect may lead to fundamentally different structures being produced.

It is likely that humic formation is a dynamic process with new materials being added to the polymer over time, while other materials are being lost as a result of microbial attack on the humics. The bulk polymer may be the result of previous condensation reactions, decomposition of biopolymers, or quite likely both. This means that at different times a humic acid, even at a given site, will be different. It is also clear that aging in the soil affects the nature of the precursors that may be incorporated into the polymers (Nelson et al. 1979), and that the humic polymer itself changes with age as discussed by Martin and Haider (1980b), and that this aging may extend over a very long period of time (Schnitzer and Calderoni 1985). The many studies that have been conducted on the formation of model humics and their reaction with a variety of soil organic materials has played a major role, along with great strides in analytical chemistry, in advancing our understanding of how humic acids may form in nature and how humics react with other types of molecules, including plant constituents and pollutants. It has also allowed us to gain a basic understanding of the components that may come together to make humic acid. The use of controlled model ecoystems can be envisioned as a logical extension of this work with chemically synthesized model humic acids. We are moving in the direction of a system that is somewhat closer to nature in its basic composition and that can be manipulated during the process of humic formation in ways that will permit a greater understanding of the processes involved. In another sense, controlled model ecosystems can be used to study degradative processes that may transform biopolymers into humic materials. It is possible to construct a model system which makes no assumptions as to what mechanisms are involved.

SUGGESTED CONTROLLED MODEL ECOSYSTEMS

The goal of generating controlled model ecosystems is to develop a system that allows a well characterized and representative humic material to be produced by a methodology that incorporates the kind of reactions that might proceed in nature, and also gives the researcher a high degree of control over the reactions that might proceed in the ecosystem. One of the new approaches that could be built into controlled model ecosystems is the ability to sample over time as the humic acid formation reactions proceed. Being able to follow the progress of the reactions can reveal a great deal about the mechanisms involved. In the sections that follow I suggest three new approaches for generating humic acids in model ecosystems of varying degrees of complexity. There are certainly additional approaches that could be useful. A general suggestion that could apply to all of them is to use radiolabeled materials (either biopolymers or individual compounds) whenever possible. The use of radiolabeled compounds makes conducting mass balances on the reactions significantly easier and makes identification of products simpler.

Sequential additions of reactants

Natural ecosystems where humic acids are formed contain a complex mixture of chemical precursors, as well as biotic and abiotic catalysts. Most studies on the formation of model humic acids have relied on either individual compounds or mixtures of relatively similar compounds, but with a single catalyst. A logical extension of this research would be to construct a model ecosystem composed of a natural soil, with its indigenous population of organisms, to which we can add any specific chemical or bioploymer precursors we might choose. To prejudice the system in favor of the formation of humics at a somewhat more rapid rate than might occur in nature, various oxidative catalysts, or specific microbes can be added, either individually, together, or in some kind of sequential pattern.

Since it is desirable to follow the progress of humic formation over time, it will be useful in these systems to regulate the rate of formation. Most model humic formation studies have involved fairly high concentrations of reactants and short incubation times, from 24 hours to as long as a few days. In the model ecosystems it is desirable to follow the reaction over a period of many months. Several options are available for controlling the rate of reaction. In all cases the options available would also produce conditions closer to what exists in nature. These options include allowing the formation to proceed at neutral pH rather than the alkaline pH commonly used. It is certain that humic formation proceeds at neutral pH but will go at a somewhat slower rate. Other options include varying the concentration of the chemical precursors or biopolymers. This would slow down the rate of reaction at low concentrations or accelerate it at higher levels. An additional option would be to regulate the oxygen levels, which has been shown to influence the rate of reaction with some catalysts.

These studies could be conducted in a fairly large volume container, with the removal of samples over time for characterization by any of the large number of isolation, separation, and characterization techniques available.

Soil column study

A controlled model ecosystem could be constructed using a fairly large soil column through which a solution containing any of a variety of biopolymer or specific chemical precursors, catalysts, or any mixture of materials desired could be passed. The solution could be collected after passage through the column and analyzed over time. It is also possible to have a recycle system in which the liquid collected at the bottom recycles through the top and passes through the soil column continuously. This type of perfusion apparatus has been used extensively in soil degradation studies.

The soil columns could be seeded with a natural soil and its associated microorganisms or some synthetic support, such as glass beads, sand, or ground coal. If a synthetic support is used the column would need to be seeded with a mixture of microorganisms derived from soil. In systems using natural soil, multiple columns could be set up simultaneously, each containing a different soil or substrate and fed different mixtures of precursor molecules, oxidative catalysts, or physical/chemical additions. In this type of system the feed solution going through the column could be adjusted to contain whatever precursor substances, hopefully in radiolabeled forms, the investigator may want to study. These are allowed to react within the soil column under the conditions imposed. By monitoring the formation of materials over time it should be possible to trace the progress of humic formation under the imposed conditions.

This system has several options for collecting samples. The simplest would be to sample the materials in the effluent from the column whether operated in a recycle or once-through mode. Soil samples can be collected over time through sampling ports installed in the sides of the column, which permit cores to be collected. At the end of any given period of time either part or all of the contents of the column can be removed and subjected to any isolation or fractionation procedures desired by the researchers.

As with the sequential addition technique discussed above, it is possible to control the rate of reaction either by altering the concentration of the precursors in the feed, the pH of the system, or the oxygen content. It is also possible to have various labeled precursors added together or in some sequential fashion. The incorporation of each of these precursors into the polymer could then be followed. Alternatively, the degradation of individual or mixed biopolymers could be followed over time under a variety of imposed conditions. Since the system could contain a natural soil microbial community it should be possible to simulate the reactions these organisms would carry out in soil.

This soil column system offers the advantage of being a relatively good simulation of soil systems, is simple to build and operate, and offers considerable flexibility in terms of the sequence and manner in which materials can be provided to the system. A major disadvantage of this approach may result from the polymeric material formed being retained in the column instead of passing through in the effluent. This would make the sampling somewhat more difficult and may require that multiple columns be constructed that are sacrificed after various periods of time. An additional problem may be the adsorption of added precursors onto the soil matrix in a form that is unavailable for reaction to form the polymer. This is a situation not greatly different from what occurs in soil, but it may mean that extended periods of time will be required before humic polymers are produced in large quantities.

Chemostat studies

For many years chemostats have been used as a basic culture device in many areas of microbiology (Tempest 1970; Veldkamp 1977). They represent a means by which very stable, continuous cultures can be maintained under a variety of imposed conditions. The biomass of the microbial community in the chemostat can be regulated by the concentration of utilizable nutrients in the feed solution. The growth, and potentially the reaction rate of the organisms, can be controlled by the rate at which nutrients are introduced into the system. It is also possible, by use of vigorous mixing, to maintain suspended particulate material within the chemostat. The structure of the chemostats ranges from a simple flask with an outlet somewhere in the side, that allows the volume to remain constant, to complex, commerically available devices which have provision for mixing, aeration, pH, and temperature control. The major advantage in the use of a chemostat for the synthesis of humic acids is that a stable and relatively constant microbial population can be maintained to catalyze the humic acid formation. Certainly, for most studies a very simple chemostat would probably be quite adequate.

The chemostat to be used could be inoculated with either a suspension of soil or a mixture of organisms derived from soil. In either case the seed would be placed in the chemostat and fed a mixture of substrates to allow the growth of the organisms. In most cases it has been found advisable to feed the organisms at a fairly low dilution rate initially to allow a population of organisms to become established. Subsequently, the rate of feeding can be changed to select for rapid growing organisms by use of a high dilution rate or slow-growing organisms by a low dilution rate. As an alternative, pure cultures of known organisms that synthesize or degrade humic precursors could be used.

The researcher has a great deal of latitude in the provision of constituents in the solution fed to the chemostat. In addition to the nutrients required to support growth of the organisms, various known or suspected precursors of humic acids could be included. Oxidative enzymes or inorganic catalysts could also be added to the feed solution. Once the chemostat has stabilized, both the composition of the feed and concentration of various constituents could be altered over time and the impact on the formation of humic polymers measured. The major advantage of the chemostat is that we have a relatively stable and constant catalytic system to which various materials can be added and the impact of those additions on humic formation noted. The sampling of the system is quite simple because there is a constant production of effluent which, due to mixing, will contain a representative sample of what is in the reactor. The effluent samples can be subjected to isolation, fractionation, or any other analytical regime that is necessary. The amount of material produced as effluent is directly proportional to the size of the chemostat and the rate at which it is being fed. If a large quantity of effluent is required then either a large volume chemostat or a rapid dilution rate should be used. Since the humic formation reactions are likely to be quite slow in a system such as this, low dilution rates would be advisable. If too high a rate of feeding is used then, due to dilution, the organisms and/ or catalytic systems in the reactor have the chance to wash out more rapidly than they can be synthesized.

One of the major advantages of the chemostat is the degree of control that the experimenter has over the conditions within the unit. In addtion to controlling the concentration of the various biological and chemical reactants within the chemostat, the rate of reaction can also be controlled. Specific organisms can be selected for either by altering the nature of the feed itself or by the addition or deletion of specific chemical requirements. The control of pH, temperature, and aeration are all very easily accomplished. Another major advantage is that the chemostat can be operated for periods as long as the investigator may desire. This can include times as long as several years. Once the conditions for humic acid formation are optimized the researcher would have a constant and stable source of material for experimentation.

There are certainly other approaches to forming controlled model ecosystems that could potentially be applied to the study of humic acid formation. However, the ones suggested above appear to be those that could be most readily applied to the study of humic acids and would yield the kind of information that appears to be necessary to provide answers to the many questions that are the focus of this conference.

REFERENCES

Bondietti, E.; Martin, J.P.; and Haider, K. 1972. Stabilization of amino sugar units in humic-like polymers. *Soil Sci. Soc. Am. Proc.* **36**: 597–602.

Brannon, C.A., and Sommers, L.E. 1985. Stability and mineralization of organic phosphorus incorporated into model humic polymers. *Soil Biol. Biochem.* **17**: 221–227.

Cheshire, M.V. 1977. Origins and stability of soil polysaccharides. *J. Soil Sci.* **28**: 1–10.

Filip, Z.; Haider, K.; Beutelsphacher, H.; and Martin, J.P. 1974. Comparisons of IR-spectra from melanins of microscopic soil fungi, humic acids and model phenol polymers. *Geoderma* **11**: 37–52.

Flaig, W.; Beutelspacher, H.; and Rietz, E. 1975. Chemical composition and physical properties of humic substances. In: Soil Components, ed. J.E. Gieseking, vol. 1, pp. 1–211. New York: Springer-Verlag.

Goh, K.M., and Stevenson, F.J. 1971. Comparison of infrared spectra of synthetic and natural humic and fulvic acids. *Soil Sci.* **112**: 392–400.

Haider, K.; Fredrick, L.R.; and Flaig, W. 1965. Reactions between amino acid compounds and phenols during oxidation. *Plant Soil* **22**: 49–64.

Haider, K.; Martin, J.P.; and Filip Z. 1975. Humus biochemistry. In: Soil Biochemistry, eds. E.A. Paul and A.D. McLaren, vol. 4, pp. 195–244. New York: Marcel Dekker.

Ladd, J.N., and Butler, J.H.A. 1966. Comparison of some properties of soil humic acids and synthetic phenolic polymers incorporating amino derivatives. *Aust. J. Soil Res.* **4**: 41–54.

Ladd, J.N., and Butler, J.H.A. 1975. Humus-enzyme systems and synthetic organic polymer-enzyme analogs. In: Soil Biochemistry, eds. E.A. Paul and A.D. McLaren, vol. 4, pp. 143–194. New York: Marcel Dekker.

Lalvani, S.; Pata, M.; and Couglin, R.W. 1986. Electrochemical oxidation of lignite in basic media. *Fuel* **65**: 122–128.

Liogon'kii, B.I.; Ragimov, A.V.; and Ragimov, I.I. 1985. Mechanism of polycondensation of hydroxyaromatic compounds as a model for the formation of polyconjugated humus fragments. *Dokl. Akad. Nauk. SSSR.* **282**: 1429–1433.

Martin, J.P., and Haider, K. 1980a. A comparison of the use of phenolase and peroxidase for the synthesis of model humic acid-type polymers. *Soil Sci. Soc. Am. J.* **44**: 983–988.

Martin, J.P., and Haider, K. 1980b. Microbial degradation and stabilization of ^{14}C-labeled lignins, phenols, and phenolic polymers in relation to soil humus formation. In: Lignin Biodegradation: Microbiology, Chemistry, and Potential Applications, eds. T.K. Kirk, T. Higuchi, and H.M. Chang, vol. 1, pp. 77–100. Boca Raton, FL: CRC Press.

Martin, J.P.; Haider, K.; and Bondietti, E. 1972. Properties of model humic acids synthesized by phenolase and autooxidation of phenols and other compounds formed by soil fungi. In: Proceedings of the International Meeting on Humic Substances, Nieuwersluis, eds. D. Povoledo and H.L. Golterman, pp. 171–185. Wageningen: Pudoc.

Martin, J.P.; Parsa, A.A.; and Haider, K. 1978. Influence of intimate association with humic polymers on biodegradation of [14C]labeled organic substrates in soil. *Soil Biol. Biochem.* **10**: 483–486.

Mathur, S.P., and Schnitzer, M. 1978. A chemical and spectroscopic characterization of some synthetic analogs of humic acids. *Soil Sci. Soc. Am. J.* **42**: 591–596.

Nelson, D.W.; Martin, J.P.; and Ervin, J.O. 1979. Decomposition of microbial cells and components in soil and their stabilization through complexing with model humic acid-type phenolic polymers. *Soil Sci. Soc. Am. J.* **43**: 84–88.

Saiz-Jimenez, C.; Haider, K.; and Martin, J.P. 1975. Anthraquinones and phenols as intermediates in the formation of dark-colored, humic acid-like pigments by *Eurotium echinulatum. Soil Sci. Soc. Am. Proc.* **39**: 649–653.

Schnitzer, M., and Calderoni, G. 1985. Some chemical characteristics of peleosol humic acids. *Chem. Geol.* **53**: 175–184.

Shindo, H., and Huang, P.M. 1985. The catalytic power of inorganic components in the abiotic synthesis of hydroquinone-derived humic polymers. *Appl. Clay Sci.* **1**: 71–81.

Stott, D.E., and Martin, J.P. 1987. Biochemistry of natural and synthetic polymers in soil. *Soil Sci. Soc. Am.* in press.

Tempest, D.W. 1970. The continuous cultivation of microorganisms. I. Theory of the chemostat. In: Methods of Microbiology, eds. J.R. Norris and D.W. Ribbons, vol. 2, pp. 259–276. New York: Academic Press.

Veldkamp, H. 1977. Ecological studies with a chemostat. *Adv. Micro. Ecol.* **1**: 59–94.

Wang, T.S.C.; Chen, J-H.; and Hsiang W-M. 1985. Catalytic synthesis of humic acids containing various amino acids and dipeptides. *Soil Sci.* **140**: 3–10.

Standing, left to right:
Hans-Rolf Schulten, Konrad Haider, George Harvey, Fred Pfaender

Seated (center), left to right:
John Ertel, Russ Christman, Wolfgang Flaig, Jim Martin

Seated (front), left to right:
Pat Hatcher, Peter Behmel, John Hedges

Humic Substances and Their Role in the Environment
eds. F.H. Frimmel and R.F. Christman, pp. 105–112
John Wiley & Sons Limited
© S. Bernhard, Dahlem Konferenzen, 1988.

Genesis Group Report

J.R. Ertel, Rapporteur

P. Behmel

R.F. Christman (Moderator)

W.J.A. Flaig

K.M. Haider

G.R. Harvey

P.G. Hatcher

J.I. Hedges

J.P. Martin

F.K. Pfaender

H.-R. Schulten

Abstract.   Humic substances are transients in the global carbon cycle and thus their formation cannot be considered without their destruction. With such a variety of possible precursors and environmental conditions, it is highly unlikely that a single uniform process is responsible for humic substance formation.

INTRODUCTION

The discussions of our group centered upon the present concepts concerning the precursors, pathways, and processes for the formation of humic substances in different environments. Humic substances are generally recognized as a ubiquitous component of the organic matter in soils, sediments, and fresh and marine waters. However, their role in the global carbon cycle is not always taken into consideration. As a class of compounds humic substances represent quantitatively important intermediates in the remineralization of biologically reduced carbon to CO_2.

Since humic substances represent only a portion of the organic matter produced by photosynthesis, some clarification is needed as to what organic material we will consider as humic substances. Numerous definitions exist for humic substances; the standard definition, arising from soil science research, states that fulvic acids are the base-extractable and acid-soluble portion of soil organic matter, while humic acids represent the base-soluble and acid-insoluble component. In this scheme the base-insoluble residue is called humin. Other definitions require that humic substances be the end product of a specific transformation pathway or a uniform collection of singularly formed molecules.

In order not to prejudice our concepts concerning humic substance formation, we have chosen to define humic substances operationally as having the acid-base solubility behaviors listed above. This general scheme can be applied equally well to other particle-bound humic and fulvic acids such as those found in marine sediments and riverine and oceanic suspended particles. Since there are major procedural differences in the isolation techniques between dissolved (aquatic) and particle-bound humic substances, we require that aquatic humic substances at least conform to the solubility characteristics of the particle-bound humic substances. However, major differences in isolation procedures dictate that we identify the environmental sources of humic substances with modifiers such as soil, freshwater aquatic, marine sedimentary, etc. For the purpose of discussing genesis, it is not necessary for us to detail isolation procedures; we leave that to the isolation group (see Thurman et al., this volume). However, for the sake of comparison in future research, we suggest that it is important to fully specify the isolation procedure used and to provide detailed characterizations of humic substances including elemental, functional group, and spectroscopic analyses.

Using this operational definition, humic substances are usually found to be mixtures of amorphous, acidic, colored organic compounds. Although there are many other characteristics common to humic substances, further specification of properties (such as molecular weight) will only tend to discriminate against certain environments (e.g., marine dissolved humic substances). However, it must be emphasized that the terms humic and fulvic acids imply only specific acid-base solubility behaviors and that these organic fractions isolated from diverse environments do not necessarily have similar chemical structures or result from similar source materials or formation processes.

We have chosen to discuss the genesis of humic substances separately in terms of possible natural precursors, formation pathways, and environmental influences. However, it was clear from our discussions that these 'opics are highly interwoven: certain pathways are favored under specific environmental conditions and certain precursors are more likely to react along specific formation pathways.

POSSIBLE PRECURSORS FOR HUMIC SUBSTANCE FORMATION

All biopolymers, monomers or metabolites of algae, microorganisms or vascular plants have the potential of being or becoming humic substances. The choice of potential precursors is highly dependent on the formation pathways. Since both constructive and destructive pathways are possible for genesis of humic substances, biopolymers and their degradation products are all potential precursors. The likelihood of precursors becoming humic substances depends on their reactivity under the environmental conditions present.

The biopolymer precursors include not only all of the known biological macromolecules, such as lignins, carbohydrates, proteins and triglycerides, but also heretofore uncharacterized biochemical compounds, for example the aliphatic polymers found in vascular plant cuticles. Major advances in instrumentation and techniques that permit characterization of macromolecules, such as ^{13}C NMR and analytical pyrolysis, have revealed previously undetectable biopolymers that might be incorporated into humic substances. Even relatively minor cellular components can make significant contributions to humic substances if concentrated by remineralization of more labile biochemicals. Thus the primary justification for the melanoidin hypothesis applied to the marine environment—that phytoplankton are mostly proteins and carbohydrates—is not valid if these labile biopolymers are rapidly converted to CO_2.

The direct incorporation of biopolymers into humic and fulvic acid mixtures is an underemphasized aspect of the operational definition of humic substances. Certain types of biopolymers (like uronic acids, proteins, and some lignins) have the required solubility behavior of humic substances without modification and thus must be considered humic substances. In environments where remineralization of labile biochemicals is slow, such as reducing sediments, these components might comprise a major portion of the humic substances.

Humic substances themselves are transients in the environment and are only relatively more stable than other molecules. Thus refractory biopolymers have a higher probability of becoming humic substances, assuming that they naturally have or can be transformed to have the required solubility behavior. Also, chemically or physically modified compounds that have become stabilized to remineralization by microorganisms, as might occur due to abiotic condensations of metabolites, are generally more likely to be incorporated into humic substances.

The importance of specific precursors in different environments is dependent not only on their distribution in the biosphere, but also on their relative stability in that environment. For example, lignins are not likely precursors for humic substances found in deep-sea sediments, since marine humic substances are thought to be autochthonous and phytoplankton contain no lignins. To date, lignins appear to be the only biopolymer with this degree of limited distribution. Lignins are a major component of terrestrial biomass. However, in aerobic soil environment lignins degrade, ultimately to CO_2, perhaps leaving behind minor, more refractory plant or microbial components. Thus lignins might be more significant precursors for humic substances in depositional environments like nearshore or lacustrine sediments where they are more refractory than phytoplankton-derived source materials.

POSSIBLE PATHWAYS AND PROCESSES FOR HUMIC SUBSTANCE FORMATION

Opposing theories abound for the formation of humic substances even in the well-studied soil environment. Laboratory evidence clearly indicates that synthetic mixtures of organic compounds having the properties of humic substances can be formed by constructive and destructive pathways and by biotic and abiotic processes. However, conclusive evidence that humic substances are formed by any one pathway or process in any single environment appears to be lacking. Since humic substance formation is a dynamic process, it seems likely that synthetic and degradative processes are occurring simultaneously. Thus the concept of a linear, unidirectional, stepwise pathway for humic substance formation is probably an oversimplification of a complex process. This also implies that there is no universal, unidirectional precursor-product relationship between humic and fulvic acids.

The chemical structure of biological precursors strongly influences pathways for humic substance formation. Highly refractory compounds like polyethylene biopolymers would have a high probability of forming humic substances directly or with slight modification (biopolymer degradation model) but would have little tendency for their aliphatic degradation products to spontaneously condense to form humic substances (abiotic condensation model). Proteins and carbohydrates might be too labile to persist as humic substances but their highly reactive degradation products could react rapidly with other monomers or humic substances. Lignins are perhaps unique in that, depending on the environmental conditions, they can either be converted directly to humic substances or yield individual phenols which readily condense to form humic substances.

Environmental conditions also affect the relative importance of different processes. Photochemical condensations are clearly favored more in the sea surface microlayer than in soils. Biotic processes likely dominate humic substance formation in areas of high microbial activity such as aerobic soils. Microorganisms remove labile biochemicals thereby concentrating refractory, "preformed" humic substances. However, at the same time metabolites are formed which have the potential to react biotically or abiotically with coexisting compounds, possibly making them more refractory. It is hard to imagine these processes decoupled, particularly for lignins.

USE OF MODELS

The importance of model compounds and model systems is in testing which precursors, pathways, and processes are possible, not in determining which is most probable. Model compounds need to be synthesized to evaluate specific, well-formulated hypotheses concerning humic substance formation, and these synthetic compounds should be subject to the most discriminating analytical techniques available. There appears to be little utility in attempting to make "real" humic substances in the laboratory or in synthesizing model compounds for hypotheses that cannot be critically scrutinized. A problem with model compounds is that they can be as structurally complex and difficult to characterize as natural humic substances.

The ability to eliminate potential precursors or processes by conclusive negative results is an underutilized aspect of hypothesis testing. Rather than attempting to "prove" a hypothesis, alternative hypotheses should be devised, model compounds synthesized that incorporate the cruxes of these hypotheses, and crucial experiments devised to systematically eliminate these possibilities. For example, due to the abundance of proteins and carbohydrates in phytoplankton, the condensation of sugars and amino acids into melanoidins has been suggested as a probable pathway for marine humic substance formation. Model melanoidin compounds resemble marine humic substances in many elemental and spectroscopic properties. However, when analyzed by ^{13}C NMR, melanoidins made purely from sugars and amino acids are considerably more aromatic than marine humic substances and thus melanoidins themselves are not the major source of marine humic substances.

It is unlikely that any suite of synthetic compounds or isolated biopolymers will ever fully resemble natural humic substances. However the ability to use isotopically labeled, pure biopolymers and synthetic compounds for biodegradation studies makes this approach indispensable. As analytical methods become more discriminating, particularly at the macromolecular level, model compounds might become more valuable for completely identifying subunits of humic substances.

Model systems have the capability of simplifying natural systems. For example, degradation studies can be performed using simple microbial communities or single isotopically labeled substrates. It is clear that such a drastic perturbation of natural system cannot yield realistic kinetic information, but it can yield comparative information on different environmental parameters.

Specific microenvironments, for example vascular plant fragments in soils or sediments, are ideal limiting situations to study specific formation processes. These endmember environments are the bridge between model systems and complex natural systems and should be used to examine the formation of humic substances in natural, though special, environments.

ENVIRONMENTAL MODULATION

As discussed previously, environments clearly influence humic substance formation not only by dictating the range of possible precursors but by influencing the type of possible processes. We do not appear able at the present time to predict structural characteristics based on environmental conditions, except those based on the selective distribution of precursors, i.e., higher aromaticity due to lignins in terrestrial environments and higher sulfur content in marine humic substances from reducing sediments.

A major problem in deciphering environmental influences is that humic substances are often isolated from different environments than those in which they were formed. For example, riverine aquatic humic substances represent, in part, the most soluble component of soil organic matter; coastal marine sedimentary humic substances as well as dissolved marine humic substances contain unambiguous terrestrial components. In addition, humic substances can be selectively transported through soils, further fractionated by interaction with riverine suspended particles and selectively removed in estuaries. This environmental processing complicates our ability to establish precursor-product relationships and to define formation pathways.

A largely unrecognized aspect of the geochemistry of humic substances is that there is generally a physical link between different types of environments. For example, humic substances are transported from soils through groundwater to rivers through estuaries to the oceans while researchers tend to concentrate only on their specific environments. Thus, crucial to understanding the formation process of humic substances is the understanding of the physical dynamics of the environmental system, of which humic substances are but a part.

FUTURE RESEARCH NEEDS

With recent advances in isolation and separation technology and in instrumentation for the characterization of macromolecules (such as ^{13}C NMR and analytical pyrolysis), we have begun to examine and challenge existing genesis theories. These approaches must be consistently applied to natural humic substances from a variety of different environments and to model compounds which critically evaluate currently accepted pathways and processes. We have identified several major research problems.

Unidentified macromolecular precursors. We need to analyze microorganisms and plant materials from marine and terrestrial environments for possible minor refractory biopolymers. Important examples of these are the aliphatic polymers found in algae and plant cuticles and peptidoglycans in bacteria. It is important to look for isotopic and molecular tracers to distinguish among these biopolymers and to unambiguously identify them. In addition the biosynthetic pathways should be delineated so that specifically labeled polymers can be made for model studies.

New proposed pathways. The environmental conditions, required precursors, and final products of the photochemical and hydrogen sulfide condensation reactions of polyunsaturated lipids need to be further delineated.

Model compounds need to be synthesized and critically evaluated in comparison with natural humic substances.

Enzymes for humic substance formation. Biotic processes play an important role in humic substance formation. However, few of the enzymes involved have been characterized. For example, the distribution of ligninolytic organisms in soils and aquatic environments needs to be determined and the activity and mode of action of their enzymes in the environment need to be studied.

The role of nitrogen and sulfur. The nitrogen and sulfur content of humic substances is characteristic of different environments and clearly affects the physical and chemical properties. However, much of the functionality of the nitrogen and sulfur is unknown, despite its crucial importance to existing formation theories.

The structure of larger subunits. The key to determining whether humic substances are remnant biopolymers or abiotic condensates lies in the randomness of the chemical bonds between small subunits. With milder degradation techniques and the present ability to characterize macromolecules these linkages can be preserved and investigated.

Relative rates of genesis from different precursors. Numerous environments have mixed inputs of marine and terrestrial organic matter. Even in the same environment humic substances from marine and terrestrial precursors might form by different pathways. Examining the relative rates and structures of humic substances in model systems with labeled precursors may indicate formation pathways.

[14]C residence times of molecular components of humic substances. Determination of residence times (apparent ages) for macromolecular components and simple molecules from degradation analyses will begin to answer the question concerning the temporal homogeneity of humic substance mixtures, and will provide an important test for the "core" theory of humic substances.

The geochemistry of humic substances. Humic substances must be studied in terms of their position in the global carbon cycle and in relationship to the physical processes, such as differential transport and selective adsorption, which affect them. As much emphasis needs to be placed on the geochemical questions concerning large scale processes and attending budgets as has been placed on the chemical questions. Specialists of humic substances processes in soils, rivers, and the marine environment jointly need to address the important questions at the interfaces of their specific environments.

REFERENCES

Aiken, G.R.; McKnight, D.M.; Wershaw, R.L.; and MacCarthy, P., eds. 1985. Humic Substances in Soils, Sediment and Water. New York: John Wiley & Sons.

Ertel, J.R., and Hedges, J.I. 1985. Sources of sedimentary humic substances: vascular plant debris. *Geochim. Cosmochim. Acta.* **28**: 1523–1535.

Hatcher, P.G.; Spiker, E.C.; Szeverenyi, N.M.; and Maciel, G.E. 1983. Selective preservation and origin of petroleum-forming aquatic kerogen. *Nature* **305**: 498–501.

Nip, M.; Tegelaar, E.W.; de Leeuw, J.W.; Schenck, P.A.; and Holloway, P.J. 1986. A new non-saponifiable highly aliphatic and resistant biopolymer in plant cuticles. *Naturwissenschaften* **73**: 579–585.

Platt, J.R. 1964. Strong inference. *Science* **146**: 347–353.

Humic Substances and Their Role in the Environment
eds. F.H. Frimmel and R.F. Christman, pp. 113–132
John Wiley & Sons Limited
© S. Bernhard, Dahlem Konferenzen, 1988.

Evolution of Structural Models from Consideration of Physical and Chemical Properties

W. Ziechmann

Abteilung Chemie und Biochemie im System Boden
Institut für Bodenwissenschaften der Georg-August-Universität
3400 Göttingen, F.R. Germany

Abstract. For humic substances, chemical and physical facts should be relevant to the development of structural models. The following points reflect several historical phases in their formulation:

1) fixation of concepts and an extensive nomenclature (1780–1850);
2) presentation of the first detailed structural formula (1920–1970);
3) comparable endeavors to derive models without chemical details (1950–1960);
4) application of modern physical techniques for a refinement of structural understanding (1950); and
5) analysis of the genesis of humic substances as a contribution to their structural problem.

What these efforts have in common is that new insights should promote a better understanding of these natural materials and their activity in the environment.

If only we knew exactly what we cannot know about humic substances on principle, we could know more about them.

INTRODUCTION

The exploding development of natural sciences has inevitably brought about the necessity of introducing order in recently acquired knowledge. An example of this ordering process is the introduction of the periodic system as a synopsis of observations and experiences in the field of inorganic chemistry in the nineteenth century.

It seems that the chemistry of humic substances after its initial stage, including manifold experiments, is reaching a similar situation. Therefore it is desirable to discuss possible models and concepts involving structural aspects for these natural substances.

Presuppositions

The structural models for humic substances should be derived from their physical and chemical properties, and therefore humic substances should also have a reproducible, comprehensive, and consistent structure. But have humic substances valid formulae for all particles of a fraction?

Aims

As in all comparable cases we must evolve extensive, yet practical, and schematic models to pull together the manifold experiences involving humic substances. Beyond that, a model—in whatever way it is formulated—must include the possibility of understanding the relations between the structure and quality and the chemical properties and activities of humic substances.

Criteria

Under these circumstances criteria for the validity of a model should be oriented around two facts: explanation and prediction (Fig. 1). Explanation means a sensible derivation of physical and chemical qualities from the model and the fixation of the model by these qualities (see Fig. 1(1)). Prediction means the estimation of unknown properties by an interpretation of the model (see Fig. 1(2)).

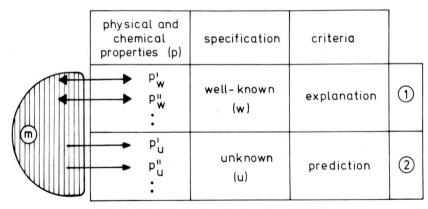

physical and chemical properties (p)	specification	criteria	
p'_w p''_w ⋮	well-known (w)	explanation	①
p'_u p''_u ⋮	unknown (u)	prediction	②

Fig. 1—Functions of a structural model (m).

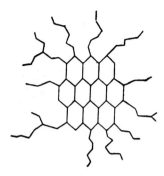

Fig. 2—A structural model after Kasatochkin.

THE BEGINNING OF THE FORMULATION OF STRUCTURAL MODELS OF HUMIC SUBSTANCES

Clearly a process such as the evolution of structural models for humic substances has extended over a long time period; even today it is not yet completed. Perhaps the essential advantage is this: we see some problems clearer now, but still are a considerable distance from a definite solution. The start of this development may be seen in the first attempts of classical chemistry to treat and characterize humic substances such as humic acids, fulvic acids, hymatomelanic acids, and even crenic or apocrenic acids, etc. These attempts were connected with such names as Berzelius (1839), Hoppe-Seyler (1889), Oden, Springer, and others. Perhaps a far-reaching misunderstanding has evolved from these attempts to consider and separate humic substances like definable and pure chemical compounds.

The next point in this direction must be the assignment of a structural formula of humic substances. Some examples may help support this fact.

Kasatochkin's model (1951) shows a nucleus of six condensed carbon rings with aliphatic side chains (see Fig. 2). In this model functional groups and other details of a humic particle are not yet considered and therefore only its structural framework or skeleton stands in the foreground.

We must query, however, the shape of the molecule in this model because of experiments which have proven that humic particles are spheric. In regards to the formula mentioned above this fact is doubtful because the condensed rings must imply either a plane system, if the rings were aromatic, or a wave-shaped system, in the case where the rings are aliphatic, like those of diamonds.

Dragunov et al. (1948) offered a model with aromatic and quinonoid rings connected by several bridges (see Fig. 3). The observed functional groups of humic acids are attached like nonhumic substances (e.g., carbohydrates, peptides).

In principle, Kleinhempel's model (1970) is only slightly altered but offers more details (see Fig. 4). The molecules in Schnitzer's model (1972) are

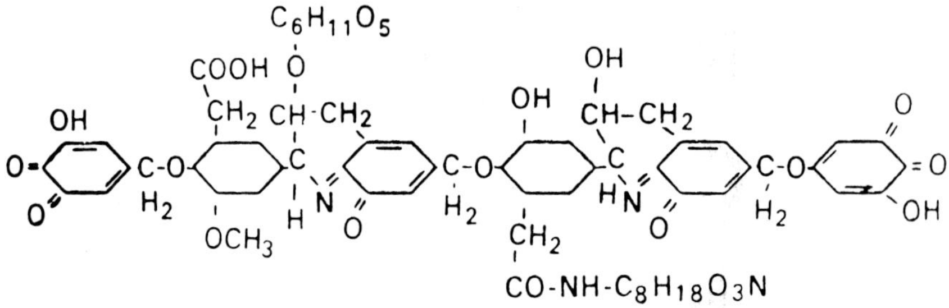

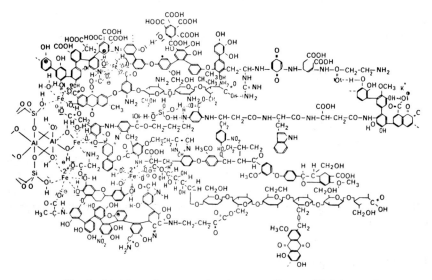

Fig. 3—Dragunov's structure of a humic acid molecule.

Fig. 4—Linking of structural units according to Kleinhempel.

composed of distinct benzene rings, substituted by phenolic and carboxylic acid groups (see Fig. 5). The interconnection of the so-called "building blocks" is based merely on hydrogen bonds.

Many experiments have previously investigated the elemental composition, functional groups, and molecular weights of humic substances. Some expectations may have been satisfied by long-favored general interpretations, which had attractive features and considerable plausibility in the field of "soil fertility" and similar structurally imprecise approaches. Since most of these interpretations are obviously not based on classical structure analysis,

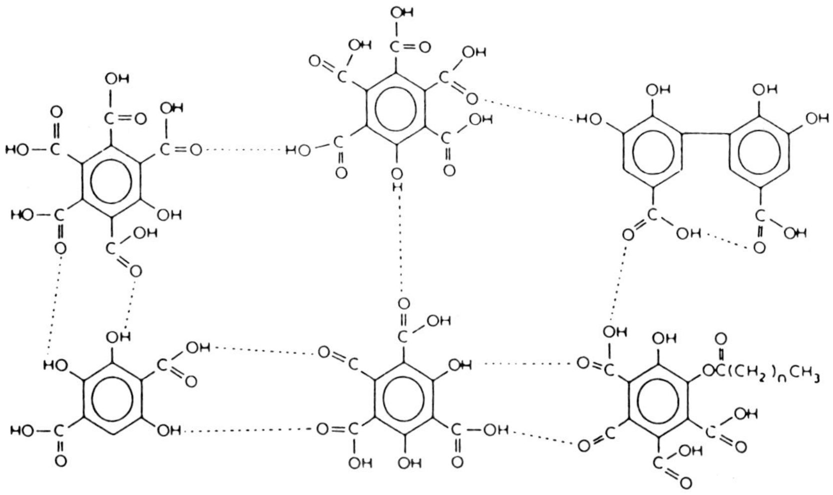

Fig. 5—The structure of fulvic acids according to Schnitzer.

using the methods of organic chemistry, they must be viewed as a first approximation to a "fictive or pseudo problem."[1]

Obviously it is not always one's intention to construct an authentic copy of a particle of a humic acid but rather to mediate an approximate reproduction from its arrangement. On the other hand, structural formulae such as these are already a statement in themselves and follow a specific interpretation, according to basic chemistry.

FURTHER DEVELOPMENTS

After this period the next step was toward the formulation of structural models which lack distinctive details. The improvement of these models led to an apparent recession. Nevertheless, with the abandonment of chemical details the model could reach yet a greater content for information. An attempt in this direction is the formulation of the principle of structure of humic substances by Thiele and Kettner (1953), which involved a coordination with parts of molecules and their structural functions (see Fig. 6).

In this connection some parts of molecules are granted a distinct function in the system. The model was modified by Pauli (1967), who considered the spatial order of magnitude (see Fig. 7).

[1]A criterion for the validity of these structural models is their transferability to molecular models like Stuart skull-caps, Dreiding or others.

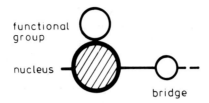

Fig. 6—The structural model of humic substances by Thiele and Kettner.

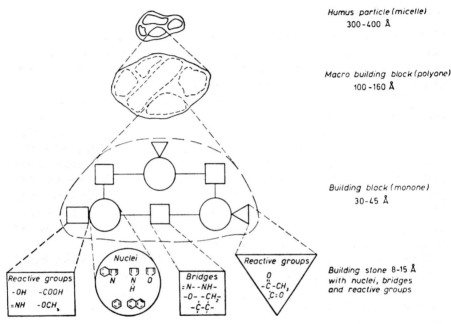

Fig. 7—The structure of humus molecules according to Pauli.

Pauli describes the humic particle as a micelle, from the point of view of colloid chemistry, formed by polyions and monones. In this region of the building stones, nuclei, bridges, and functional groups can be observed in more detail and in distinct dimentions.

Cheshire et al. (1967) proposed a generalized scheme with a polycyclic aromatic nucleus and attached polypeptides, carbohydrates, phenolic acids, and metals. In addition, a lot of other models still claim to contribute a distinct aspect to the structure of humic substances. However, it was not always considered whether the method chosen may have influenced the results of experiments in the field of humic substances. For example, the determination of the molecular weight of a natural humic acid from B_h horizon of a podzol profile was made by measuring the decrease of the melting point of urea, and by the ultracentrifuge method. The value obtained by the first method indicates a molecular weight of 1,000, the second method indicates about 10,000. Referring to the first method we could assume that the hydrogen bonds were dissociated by urea, and it is very likely that fragments (the so-called "building blocks") were measured (Ziechmann 1980).

NEW METHODS — NEW IMAGINATIONS

Quite recently several authors have used new experimental concepts to develop methods for the solution of the structural problem of humic substances. That has meant that the method per se was pushed into the foreground and not even a structural formula was derived.

Thurman and Malcolm (1983) attempted in the three-step approach, "separate-degrade-identify," to confront the problem of structural analysis of humic substances. The separation methods they applied were gel filtration, adsorption chromatography, pH gradient adsorption chromatography, and anion and cation exchange. New degradation technques are the chlorination and methylation with ^{13}C reagents, and identification took place with ^{13}C nuclear magnetic resonance (^{13}C NMR). Finally, numerous data were obtained by elemental analysis and by the determination of the funcional groups, the aromatic and aliphatic parts, the amount of amino acids and carbohydrates, and the molecular weight.

Gillam and Wilson (1983) have extended the ^{13}C NMR to marine humic substances and Hatcher and co-workers (1983) to humic substances from peat, soils, shelf sediments and aquatic humins. Fluorescence phenomena were investiged by Visser (1983) on aquatic humic materials and substances of microbial cultures.

Degradation of organic substances with uniform structure is in many cases a suitable method for an analysis of the structure of humic substances. The question arises whether any success can be expected by this treatment, because humic substances consists of a nonhomogeneous material and their fragments are very reactive, suitable enough to form artifacts. The following methods are known and practiced in the treatment of humic substances:

1) hydrolysis with acids and bases;
2) oxidation with $KMnO_4$, H_2O_2, CuO-$NaOH$, nitrobenzene, etc.;
3) reduction with Zn dust, Na amalgam;
4) pyrolysis;
5) depolymerization; and
6) microbiological degradation.

Through these methods a great number of defined, low-molecular organic fragments have been obtained, and many of these fragments may be genuine monomers of humic materials. We must ask, however, whether it is possible to reproduce the original molecule, or at least its basic structure from this scattered mosaic.

In contrast to lignins (Nimz 1974) neither a leading motive (leitmotiv) nor selected substances are dominant in humification (e.g., coniferylic alcohol, etc.). Apart from this problem it is not evident whether the fragments investigated are artificial products or not. The application of alkaline solvents (NaOH) definitely leads to altered substances.

SOME REMARKS ABOUT THE GENESIS OF HUMIC SUBSTANCES AND THE CONSEQUENCES FOR A PATTERN OF THEIR BASIC STRUCTURE

New analytical techniques reveal new possibilities for the discussion concerning the structure of humic substances. The genesis of natural humic substances or the analysis of their synthetic variants as models also gives information leading to a certain understanding of their natural components. Here the substances, as well as their origin, should be kept in mind. The exceptional status of humic substances is attributable to their genesis and the special conditions of their formation. Our knowledge, so far, is represented in Fig. 8. The most important reactions involved in formation are:

1) the decomposition of carbohydrates and other natural nonaromatic products, and
2) the synthesis derived from aromatics, like phenols, and the decomposition products of lignins and other aromatic products.[2]

With regard to Fig. 8 (1), the initial phase of the nearly ubiquitous process of the alteration of organic material in soils to the stable intermediate state of humic substances (hs) is determined by soil organisms and microbes. After a degradation of the different materials (1.1, Fig. 8), the production of precursors and the formation of genuine humic substances takes place (1.2, Fig. 8). It is difficult to find a typical reaction mechanism for this nonuniform process of humification and it seems that the Maillard reaction could readily prove this assumption.

[2] It is worth noting that colored substances are formed under conditions of the early atmosphere by reaction of H_2O, CH_4, NH_3, and H_2 after amino acids, carbohydrates, adenin, and other defined substances have originated (Miller 1953). The conditions for the initial phase of this experiment are adequate as a state of matter, which can be characterized as "chaos." In addition to this report, there are other indications which cite humic substances as a result of reactions in a state of chaos: each particle is able to react with every other one without any dominating type or order of reaction. The chemical equivalent for this situation is the existence of radicals.

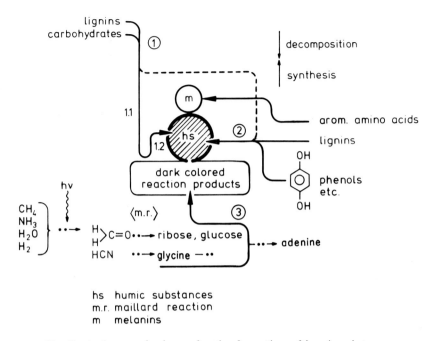

hs humic substances
m.r. maillard reaction
m   melanins

Fig. 8—A shortened scheme for the formation of humic substances.

With regard to Fig. 8 (2), aromatic acids, phenols, and many defined low molecular compounds have to be taken into account as starting material for the secondary aromatics of lignins and their fragments.

The initial phase of the reaction, by which humic substances are obtained from distinct aromatic substances, is actively investigated. Suggested mechanisms are based on electron transitions with radical states in an earlier phase. For the formation of humic substances from initial aromatic substances it is attractive to assume five phases:

1) biogenesis of aromatic or nonaromatic compounds,
2) microbiological decomposition of aromatics,
3) formation of radicals,
4) phase of conformation, and
5) formation of a system of humic substances.

Model reactions with hydroquinone and phenols verify the third phase of the process through the observation of free radicals by ESR during the humification of natural substances. The interruption of the humification (autoxidation) of hydroquinone by addition of the radical ·NO leads to the products shown below (Ziechmann 1980):

Table 1   Humic substances and reactants of the environment.

Reactants	State of shs	Humic Substance	Kind of Linkage (dominating)
Carbohydrates (Hasselmann, in press)	1	pha	cov
Amino acids (peptides) (Müller-Wegener 1982; Niemeyer 1984)	1–2	pha	ε-dac→cov
Phenols (Müller-Wegener 1976)	1 2	pha ha	cov ε-dac
Herbicides (s-triazines) (Müller-Wegener 1977)	2	ha	ε-dac→cov
Steroids (Ziechmann, unpublished)	2	ha	ε-dac
Hydrocarbons (Kress 1977)	2	ha	ε-dac
Clay minerals (Ziechmann 1980)	1,2	pha ha	$\dfrac{\text{cov}}{\epsilon\text{-dac}}\Big\}\rightarrow\dfrac{\text{occlusion}}{\text{compounds}}$
Enzymes (Ziechmann 1980)	2	ha	cov/ε-dac

pha	precursors of humic acids
ha	humic acids
cov	covalent bonds
shs	system of humic substances
1	system in state 1 : in "statu nascendi"
2	system in state 2 : in stationary state
ε-dac	
ε donor-acceptor complexes	

The findings of Diebler and co-workers (1961), regarding the formation of a semiquinonoid radical, are further proof of this presumption. A relatively small content of radicals in the humic acids is an indication for their stability caused by several mesomeric states.

The fourth step is confirmed by the proof of ϵ donor-acceptor relations and the investigation of complexes between humic and important nonhumic substances, or other humic material (see Table 1). A special, but comparable case is represented in the genesis of melanins starting from tyrosine. Some authors regard this group of compounds as the humic substances in the animal kingdom.

A MATHEMATICAL MODEL FOR HUMIFICATION

The synthesis of humic substances by autoxidation of hydroquinone is accomplished in the initial phase, in comparison to the genesis of natural humic substances in soils. Such a model for the humification process is supported by a remarkable conformity between the final products of both branches. The reasons is evident: in both cases radicals are formed and this fact has the consequence that humic substances cannot be uniform materials with a distinct structure.

The attempt to describe this process in mathematical terms is connected to the assumption that temporal fixed and anlytically realizable states are only dependent on the previous state. Herewith, the mathematical conception of a Markoff process is given. Markoff processes are described by distinct states (i, j) and the probabilities of transitions between them. The genesis of humic substances can be interpreted as a sequence of connected Markoff processes (Kappler and Ziechmann 1969). A complete system with the fractions F_1 to F_n is given by:

$$\sum_{j=1}^{n} p_{ij}(t_{\nu-1}) = 1, \qquad p_{ij}(t_{\nu-1}) \geqq 0$$

where $p_{ij}(t_{\nu-1})$ ($i = 1,...,_n$ and $j = 1,...,_n$) is the mass portion of which goes from fraction F_i to F_j during the interval between $t_{\nu-1}$ and t_ν ($\nu = 1,2,3...$).

For the application of the Markoff approach, a far-reaching chemical separation of the system of humic substances, in statu nascendi, is necessary during autoxidation (Fig. 9). The experimental determination of the time-dependent alteration of the quantities of single fractions (Fig. 10) is compared to the calculated data and the probabilities of transitions of the system with the four fractions:

$$F_1 \rightarrow F_2 \rightleftarrows F_3 \rightarrow F_4.$$

The question which now arises is whether or not a function exists which describes the development of a system of humic substances in agreement with the experimental results.

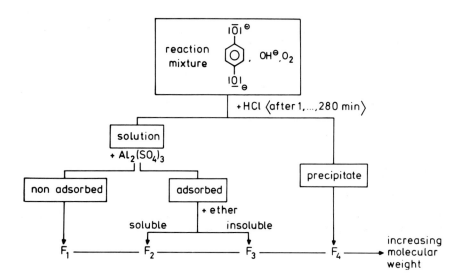

Fig. 9—Separation of a system of humic substances.

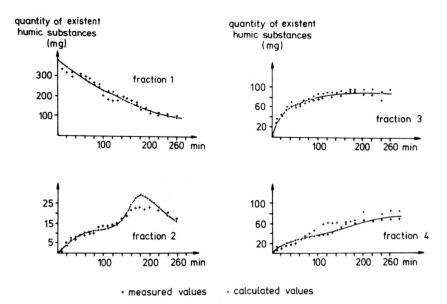

Fig. 10—Alteration of the fractions F_1 .... F_4 with time.

The H function, as the negative sum of the products of probability of the transition (i.e., the relative mass portion, p_i, of the fraction i) and its logarithm, satisfies the conditions.

$$H = - \sum p_i \cdot \log p_i, \qquad p_i \geqq 0 \ \text{ and} \sum p_i = 1$$

As a consequence, a concept of a modified interpretation of the problem of the structure of humic substances can be given by an analysis of the alteration of the H function over time (Fig. 11 (1)).

CONCLUSIONS

According to the experiments and their mathematical treatment, the curve for the alteration of the H values, over time (Fig. 11), shows a steep rise at the beginning and finally a horizontal branch. This can be interpreted as follows:

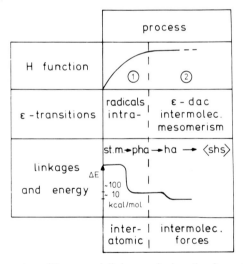

Fig. 11—H function, ε-transitions, and linkages during the formation of systems of humic substances (shs).

st.m.	starting material
pha	precursors of humic acids
ha	humic acids
ε-dac	ε donor-acceptor-complexes

| states of the system: | 1 | in statu nascendi |
| | 2 | in stationary state |

1) If the masses of humic substances are concentrated in one fraction (still no system!), the H function will disappear. A maximum is reached if the masses of a system of humic substances are uniformly distributed above all fractions as much as possible:

$$p_i = \frac{1}{n} : H = - \sum_1^n \frac{1}{n} \cdot \log \frac{1}{n}, \text{ and } = - \log \frac{1}{n} \text{ or } = \log n$$

2) There are two phases given for the development of a system of humic substances. In the first, the material is in the state of the formation of the system ("in statu nascendi"). In the second phase the system has reached its stationary state:

(a) the system in statu nascendi: $\lim_{t \to \infty} \dfrac{dH}{dt} \neq 0$

and

(b) the system in the stationary state: $\lim_{t \to \infty} \dfrac{dH}{dt} = 0$

In the first phase the genesis of the system is caused by radical reactions. Only through such reactions can substances be obtained with the qualities described. Finally, the H function asymptotically approaches a horizontal line. The rise of the curve in the first phase reflects a great shifting of material and the masses of the particles. This fact is caused by reactions of radicals or precursors of humic acids (interatomic forces) which results in a remarkable alteration of the weights and dimensions of the particles by formation of covalencies. The asymptotic shape in the second phase results from the change of intermolecular forces (humic acids) without the above-mentioned consequences: there is no substantial modification of the magnitude of particles. These typical functions for the single phases involve other substances, ultimately, the generation of a "system of humic substances." After these preliminary conclusions, some definitions may be suggested which are valid for further consideration.

Humification: formation of humic substances from different monomers as a consequence of *accidental events* (Markoff processes!). The course can be described by the H function:

$$\lim_{t \to \infty} dH/dt \neq 0$$

Humic substances represent that state of matter which follows immediately after the "chaos" and transfers over to a beginning order or preorder. Humic substances, with other materials, are the *constitutive units* of a system of humic substances.

System of humic substances: intermolecular forces cause an effect between single particles. Along with nonhumic substances, higher organized systems are formed which are the *reactive units* of humic substances. Nucleus and peripheric regions can be distinguished. The system is described by

$$\lim_{t \to \infty} dH/dt = 0$$

From these facts a pattern of the basic structures (no structural formula!) of humic substances can be derived as a structural model of these natural products (see Fig. 12).

We must distinguish between (a) a primary pattern expressed by the key words spheric material, radicals, interatomic forces (covalencies), and H function (Fig. 11 (1)) and (b) a secondary pattern with the key words intermolecular forces, ϵ donor-acceptor relations, and H function (fig. 11(1)).

This arrangement of partial structures of humic substances is based on distinct linkages, and that means specific states of electrons (Ziechmann and Kress 1977).

The following states can be distinguished:

1) covalent bond $\geqslant C:C \leqslant$, $\geqslant C:O_{\backslash}$, etc.
2) ionic bond $\geqslant C - \ddot{O}: \ominus$

3) lone electron pair $> \ddot{O}:$, $\geqslant N:$
4) unpaired electrons $\geqslant C \cdot$, $> O \cdot$ (radicals)

5) triplets (phosphorescence) $\uparrow \uparrow$, $\uparrow$ $\uparrow$
6) ϵ donor-acceptor relations

$$\langle 6\pi \rangle \Big| \langle 5\pi \rangle^{\oplus}$$
$$\langle 4\pi \rangle \Big| \langle 5\pi \rangle^{\ominus}$$

Radical formation was investigated by heating and irradiating a synthetic humic acid sample. The temperature was varied between 100° K and 373° K. The radient energy was a 1000 Watt mercury vapor lamp. At room temperature and without irradiation a low concentration of radicals (mesomeric stabilization) was obtained. At room temperature and with irradiation photolysis, a formation of π-radicals and triplet states took place. The interruption of irradiation resulted in covalent bonds. At high temperature (360° K) and without irradiation an increase of π-radicals was observed, while covalent bonds were split.

The reaction conditions were altered ca. ninety times in order to reach an equilibrium and to test the reversibility of the reactions. To prolong these relations in a third axis a criterion was added for the formation of inonic linkages, ϵ donor-acceptor complexes (ϵ-dac) and covalent bonds (Fig. 13). Ionic bonds are dominating if the relative difference between the energy of

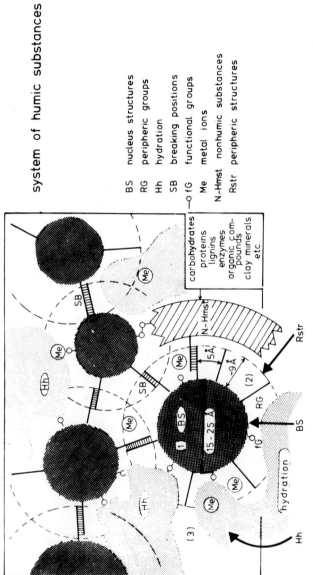

system of humic substances

BS   nucleus structures
RG   peripheric groups
Hh   hydration
SB   breaking positions
—O fG   functional groups
Me   metal ions
N.-Hmst nonhumic substances
Rstr   peripheric structures

carbohydrates
proteins
lignins
enzymes
organic com-
pounds
clay minerals
etc.

Fig. 12—Pattern of the basic structures of humic substances.

ionization I_E of the element a and the electron affinity of the element b has a great negative value, Δ_I:

$$\Delta_I = |\ I_E^a| - |\ E_A^b|.$$

Covalent bonds are formed if the difference Δ_c approaches zero and $\Delta_{\epsilon\text{-dac}}$ lies between Δ_I and Δ_c:

$$\Delta_c < \Delta_{\epsilon\text{-dac}} < \Delta_I.$$

In principle the scheme shown in Fig. 13 demonstrates these relations and is the basis for the pattern of the fundamental structures of humic substances (Fig. 12).

There are two final questions to be discussed:

1) What proofs of indication are given for the developed concept in Fig. 1(1)?
2) What characteristics of these natural substances are deducible in Fig. 1(2) and which facts on principle cannot be elucidated?

Humic substances are compounds with heterogeneous monomers and heterogenous kinds of linkages formed by random events. Therefore it is impossible to define a reproducible structure and also an exact structural formula. This simple fact is the consequence of the multiplicity of starting material and radical reactions in the first phase of humification. The donation of electrons, caused by ϵ acceptors (O_2, $Fe^{3\oplus}$, clay minerals) at the beginning of the humification, and the radical phase leads to a stabilization of the material by formation of covalent linkages with a bonding energy in the order of magnitude of 400 KJ/mol (primary pattern, Fig. 11). In the state which follows, the modalities of bonding are altered and the gain in stability is about 4–40 KJ/mol—not very high, but sufficient for a further stabilization and the formation of a system of humic substances (secondary pattern). All in all, thermodynamical processes are the naturally determining factors. factors.

There are many indications that processes of a charge transfer and the consequence of formation of ϵ donor-acceptor-complexes are typical for a system of humic substances (Ziechmann 1980). The proof of this kind of bond can be established by:

1) spectroscopic investigations (a new maximum in the adsorption spectra of the complex compared with single substances);
2) electric conductivity;
3) thin layer chromatography;
4) analysis of triplet states (Kress and Ziechmann 1977); and
5) chemical methods using reactions of defined ϵ donors or acceptors with the corresponding component of humic substances.

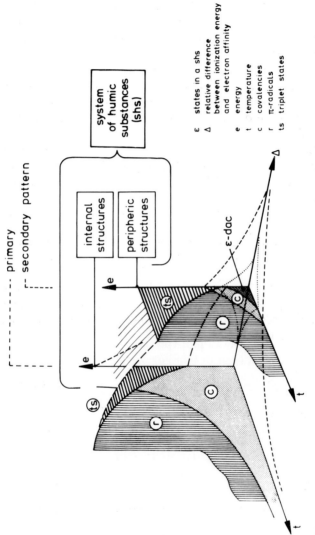

Fig. 13—States of electrons in a system of humic substances.

But what information is deducible from this model (Fig. 1(2))? Only two arguments (among others) will be mentioned here.

First, humic substances conduct an electric current in the solid state (Lentz and Ziechmann 1968; Hütten and Ziechmann 1986). Their electric conductivity is about $10^{-13}-10^{-14}\Omega^{-1}.cm^{-1}$ (weak semiconductors) in comparison with PVC $10^{-11}-10^{-16}$, pyrene 10^{-15}, and polyacetonitrile $10^{-11}\Omega^{-1}cm^{-1}$. This fact can be seen in connection with ϵ donor-acceptor relations described in the model above (Fig. 12).

Second, of greater importance are the possibilities for predicting chemical reactions of humic substances with substrates of the environment. The important parameters here are the reacting fraction (humic acids or their precursors), the state of the system (H function, Fig. 11), and the kind of linkages (Table 1).

The model (Figs. 11 and 12) helps us to understand that after an analysis of a system of humic substances,[3] it is possible to predict what kind of fraction could react with which substrates under formation of a certain kind of linkages, and also in which state of the system this reaction takes place. Finally we can return to the starting point of these considerations:

> *If only we knew exactly what we cannot know (on principle) about humic substances, we could know more about them.*

But what is impossible to elucidate in the case of humic substances? After the demonstration of the different ways of formulating a structural model for humic substances, and of estimating their meaning, some examples may be enumerated here:

1) a structural formula;
2) size and weight of particles of a fraction of the system of humic substances;
3) reaction orders for their genesis; and
4) representative elemental analysis with optimal information (like in the organic chemistry of defined compounds).

REFERENCES

Berzelius, J.J. 1839. Lehrbuch der Chemie. Dresden, Leipzig: Arnoldische Buchhandlung.

Cheshire, M.V.; Cranwell, P.A.; Falschaw, C.P.; Ford, A.J.; and Haworth, R.D. 1967. Humic acid II. Structure of humic acids. *Tetrahedron* **23**: 1667–1682.

---

[3] e.g., by the distribution pha/ha, light absorption, specific heat (Lentz and Ziechmann 1967), oxidation and consumption of O_2 (Warburg-technique), etc.

Diebler, H.; Eigen, M.; and Matthies, P. 1961. Kinetische Untersuchungen über die Bildung von p-Benzochinon aus Chinon und Hydrochinon in alkalischer Lösung. *Z. Naturforschung* **16b**: 629–637.

Dragunov, C.C.; Zhelokhovtseva, H.H.; and Strelkova, E.J. 1948. A comparative study of soil and peat humic acids. *Pochvovedenie* **7**: 409–420.

Gillam, A.H., and Wilson, M.A. 1983. Application of ^{13}C-NMR spectroscopy of the structural elucidation of dissolved marine humic substances and their chemical composition. In: Aquatic and Terrestrial Humic Materials, eds. R.F. Christman and E.T. Gjessing. Ann Arbor: Ann Arbor Science.

Hatcher, P.G.; Breger, I.A.; Dennis, L.W.; and Maciel, G.E. 1983. Solid state ^{13}C-NMR of sedimentrary humic substances: new relations on their chemical composition. In: Aquatic and Terrestrial Humic Materials, eds R.F. Christman and E.T. Gjessing. Ann Arbor: Ann Arbor Science.

Hoppe-Seyler, F. 1889. Über Huminsubstanzen, ihre Entstehung und ihre Eigenschaften. *Hoppe-Seylers Z. Physiol. Chem.* **13**: 66–122.

Hütten, U., and Ziechmann, W. 1986. Elektrische Leitfähigkeit von Huminstoffen in festem Zustand. *Mtt. Deutsch. Bodenkdl. Ges.* **45**: 93–99.

Kappler, M., and Ziechmann, W. 1969. Ein mathematisches Modell zur Beschreibung von Humifizierungsvorgängen. *Brennstoff-Chemie* **50**: 348–351.

Kasatochkin, V.I. 1951. The structure of carbonized substances. *Izv. Akad. Nauk SSSR. Otd. Tekh. Nauk* **9**.

Kleinhempel, D. 1970. Ein Beitrag zur Theorie des Huminstoffzustandes. Albrecht Thear Archives **14**: 3–14.

Kress, B.M., and Ziechmann, W. 1977. Interactions between humic substances and aromatic hydrocarbons. *Chem. Erde* **36**: 209–217.

Lentz, H., and Ziechmann, W. 1967. Physikalisch-chemische Untersuchungen an Huminstoffen. *Brennstoff-Chemie* **49**: 154–157.

Miller, S.L., 1955. Production of some organic compounds under possible primitive earth condition. *J. Am. Chem. Soc.* **77**: 2351–2361.

Müller-Wegener, U. 1976. Wechselwirkungen zwischen Phenolen und Huminstoffen. Dissertation Göttingen.

Müller-Wegener, U. 1977. Über die Bindung von S-Triazinen an Huminsäuren. *Geoderma* **19**: 227–235.

Müller-Wegener, U. 1982. Wechselwirkungen von Huminstoffen mit Aminosäuren. *Z. Pflanz. Boden.* **145**: 411–420.

Niemeyer, J. 1984. Analytische Untersuchungen zum Einbau des Stickstoffs als Strukturfaktor in Huminstoffen. Dissertation Göttingen.

Nimz, H. 1974. Das Lignin der Buche - Entwurf eines Konstitutionsschemas. *Angew. Chem.* **9**: 336–344.

Pauli, F.W. 1967. Soil Fertility. London: Adam Hilger.

Schnitzer, M. 1972. Humic substances in the environment. New York: Marcel Dekker.

Thiele, H., and Kettner, H. 1953. Über Huminsäuren. *Kolloid-Z.* **130**: 131–160.

Thurman, E.M., and Malcolm, R.L. 1983. Structural study of humic substances: new approaches and methods. In: Aquatic and Terrestrial Humic Materials, eds. R.F. Christman and E.T. Gjessing. Ann Arbor: Ann Arbor Science.

Visser, S.A. 1983. Fluorescence phenomena of humic matter of aquatic origin and microbial cultures. In: Aquatic and Terrestrial Humic Materials, eds. R.F. Christman and E.T. Gjessing. Ann Arbor: Ann Arbor Science.

Ziechmann, W. 1980. Huminstoffe. Weinheim: Verlag Chemie.

Ziechmann, W., and Kress, B. 1977. Über Elektronenzustände einer endoxydierten Huminsäure in wässriger Lösung. *Geoderma* **17**: 293–301.

Humic Substances and Their Role in the Environment
eds. F.H. Frimmel and R.F. Christman, pp. 133–148
John Wiley & Sons Limited
© S. Bernhard, Dahlem Konferenzen, 1988.

Critical Comparison of Structural Implications from Degradative and Nondegradative Approaches

D.L. Norwood

Analytical and Chemical Sciences
Research Triangle Institute
Research Triangle Park, NC 27709, U.S.A.

Abstract. This chapter describes recent examples of the degradative and nondegradative approaches to the structural analysis of humic substances. Various pitfalls in the application of each approach are presented and discussed as well as the major structural implications apparent from each. It is anticipated that the reader will gain a basic understanding of the advantages and limitations inherent in each approach.

INTRODUCTION

The study of the molecular structure of humic substances is one of the most stimulating and intriguing research areas in the field of organic geochemistry. From the very earliest reports (Dienert 1910, for example), scientists have speculated on the structures responsible for color, solubility, and other physical/chemical properties observed in humic substances. Over the past two decades, the exact nature of the molecular subunits and interatomic/ intermolecular bonds present in humic substances derived from various environmental matrices has attracted the interest of scientists from a number of applied disciplines.

Water chemists and environmental engineers are interested in the aquatic humic structures responsible for organohalide generation during drinking water disinfection with aqueous chlorine. Many of these organohalides, such as chloroform, have been shown to possess mutagenic and/or carcinogenic properties. Because of this water treatment "nuisance," the existence of aquatic humic substances has lately come to the attention of such organizations as the U.S. Environmental Protection Agency, the American Water

133

Works Association, and the World Health Organization. Water chemists and ecologists are also interested in the aquatic humic structures responsible for metal binding in streams. It appears that aquatic humic substances can play an important role in toxic metal speciation and bioavailability. Soil scientists and agricultural chemists are concerned with the structures that bind small organic molecules, such as pesticides and herbicides, to soil humic substances. The extent of terrestrial input to the structure of marine humic substances is of interest to marine scientists and oceanographers.

Two approaches to humic structure elucidation are commonly utilized. These are the so-called "degradative" and "nondegradative" approaches. Progress utilizing both of these general approaches has been directly correlated with advances in modern analytical chemical instrumentation. In the degradative approach, macromolecular humic substances are chemically or physically broken down into various subunits which are subsequently isolated and identified. Structural inferences are drawn based on the identities and yields of degradation products and the reaction processes responsible for their generation. The use of combined gas chromatography/mass spectrometry (GC/MS) for the simultaneous separation, identification, and quantification of hundreds of chemical degradation products has been essential to the development of this approach. In the nondegradative approach, isolated humic substances are analyzed directly, usually without further chemical alteration, utilizing techniques which allow structural inferences to be made. Recently developed techniques in nuclear magnetic resonance spectroscopy (NMR) have become central to this approach.

This background paper will examine each approach as it is most often applied and describe potential pitfalls inherent in each. Following each description, several major structural inferences which may be drawn from each are described. The chapter concludes by listing several new analytical techniques and briefly describing their impact on the progress of both approaches.

THE DEGRADATIVE APPROACH

Overview and examples

The degradative approach to humic structural analysis is most often applied according to the flow diagram presented in Fig. 1. First, the degradation process to be employed is selected. The particular process chosen can utilize any type of chemical reaction including oxidative, reductive, hydrolytic, and pyrolytic. The reaction process can be performed under mild or vigorous conditions as determined by the investigator. Once the reaction is complete, the reactive agent is "quenched" and the resulting degradation products

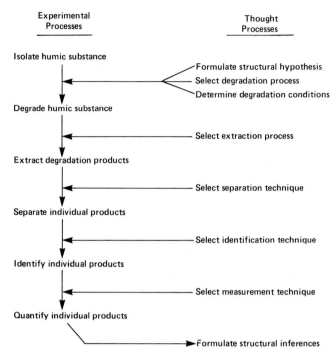

Experimental
Processes

Thought
Processes

Isolate humic substance

Formulate structural hypothesis
Select degradation process
Determine degradation conditions

Degrade humic substance

Select extraction process

Extract degradation products

Select separation technique

Separate individual products

Select identification technique

Identify individual products

Select measurement technique

Quantify individual products

Formulate structural inferences

Fig. 1—Flow diagram for application of the degradative approach to humic structural analysis.

separated from the bulk reaction mixture. The products are then individually separated, identified, and quantified.

Proper use of the scientific method as applied to the degradative approach requires that hypotheses be formulated regarding the presence of particular structural features within the humic substance under investigation. Chemical degradation experiments are then designed to attack these structural features and release degradation products indicative of their presence. Separation techniques are chosen which remove the products of interest from the bulk reaction mixture as selectively and quantitatively as possible. Individual product identifications and yield measurements which are subsequently accomplished must be definitive and accurate.

The application of chemical degradation to soil humic substances has been reviewed by Hayes and Swift (1978) and Schnitzer and Khan (1972), and to aquatic humic substances by Thurman (1985). Numerous applications of the degradative approach to soil humic substances have been accomplished by the research group headed by Morris Schnitzer of Agriculture Canada. Schnitzer and his co-workers hypothesized that aromatic nuclei were linked together in the soil humic macromolecular structure by aliphatic side chains.

Scheme I

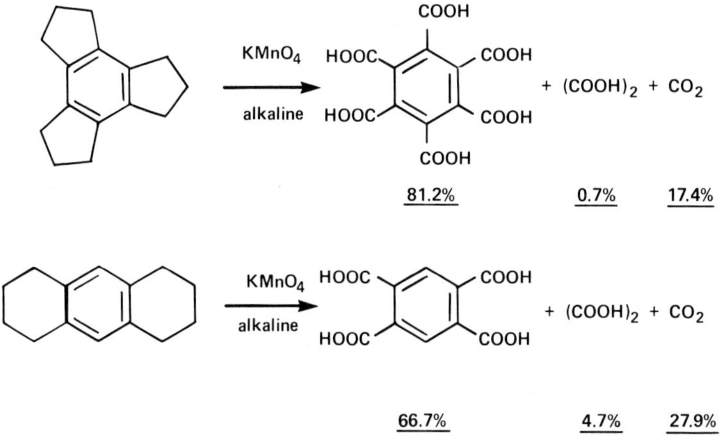

| | 81.2% | | 0.7% | 17.4% |

| | 66.7% | | 4.7% | 27.9% |

From the available scientific literature, they determined that potassium permanganate would oxidatively cleave these side chains leaving a carboxyl group at the location of each on the released aromatic nuclei. For example, Randall et al. (1938) demonstrated that the process depicted in Scheme I would occur.

Potassium permanganate oxidation of soil humic samples from numerous experiments yielded significant quantities of the expected benzene carboxylic acids in various substitution patterns, thus providing evidence in support of the original hypothesis. Based on the application of various types of chemical degradation processes (as well as nondegradative techniques) to the same soil humic samples, Schnitzer concluded that the phenolic and benzene carboxylic acids which were the usual major products "could have originated from more complex aromatic structures or could have occurred in the initial humic materials in essentially the same forms in which they were isolated but held together by relatively weak bonding" (Schnitzer 1978). If the latter hypothesis is invoked, then these aromatic nuclei can be referred to as the "building blocks" of soil humic substances. The aromatic nuclei would thus be held together by hydrogen bonds, van der Waal's forces, and π-bonding as indicated by the partial chemical structure in Fig. 2.

Schnitzer's "building block" model has been criticized in the literature, as summarized by Hayes and Swift (1978). They pointed out that the chemical degradation of soil humic substances is energetically demanding, and a structure held together for the most part by weak intermolecular forces cannot explain this fact. Further, such a structure should not display polymer properties when dissolved in neutral or basic solution. The latter fact makes it difficult to invoke this model in the case of aquatic humic substances.

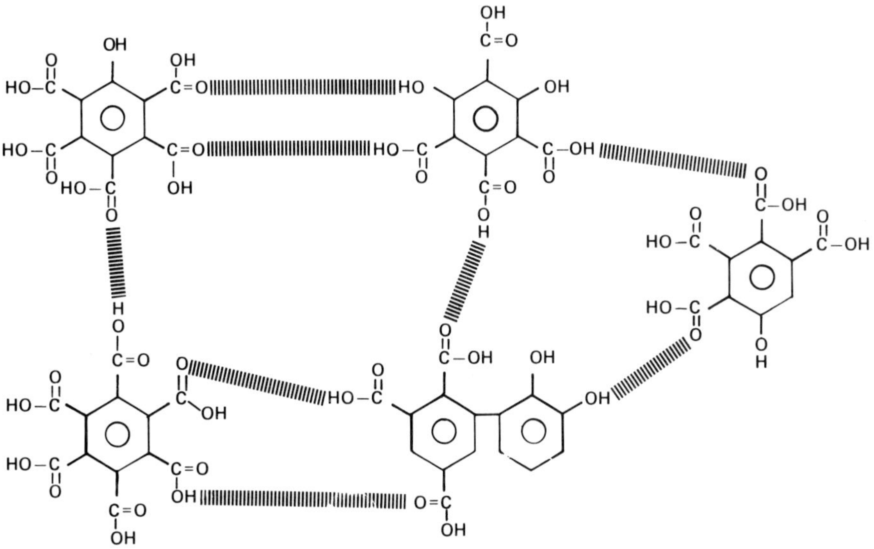

Fig. 2—"Building block" model for soil humic macromolecular structure as proposed by Schnitzer (1978).

Building on the foundation of Schnitzer's work and theories, Liao et al. (1982) undertook an investigation of the molecular structure of isolated aquatic humic substances. These workers initially hypothesized, like Schnitzer, that single aromatic nuclei were held in the humic macromolecular structure by interconnected aliphatic side chains, and proceeded to utilize potassium permanganate oxidation to cleave out these nuclei. The benzene carboxylic acids indicative of such structures were identified as important degradation products under a variety of reaction conditions. Yields of these degradation products were shown to increase with an increasing initial $KMnO_4$:C ratio, which was deemed inconsistent with the "building block" model. Further, carboxyphenylglyoxylic acids were identified as degradation products. Such structures were identified by Randall et al. (1938) from model permanganate oxidations. The reaction process shown in Scheme II is one example. This finding is highly suggestive of aliphatic side chains bonded to single aromatic nuclei in the aquatic humic macromolecular structure. Liao et al. (1982) also identified various aliphatic dicarboxylic acids as significant permanganate oxidation products. Such structures could represent the inter-aromatic linkages in the macromolecular matrix.

Christman et al. (1987) have proposed a molecular model which attempts to rationalize the potassium permanganate and base hydrolysis results of Liao et al. This hypothetical structure (which may be termed the "aromatic matrix" model) is presented in Fig. 3. The "matrix" incorporates structural

Scheme II

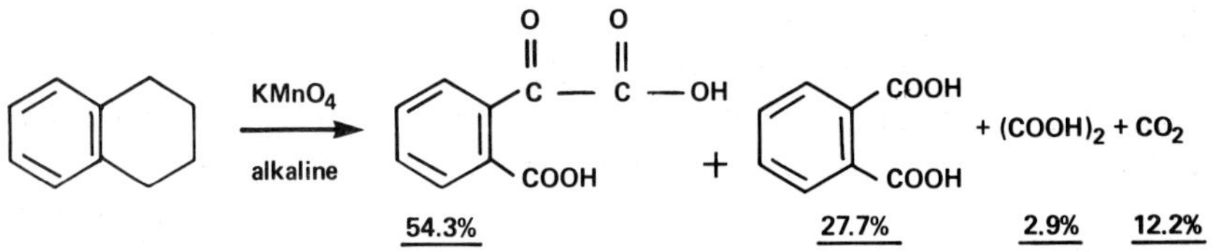

54.3%   27.7%   2.9%   12.2%

inferences based on both identities and relative yields of permanganate and base hydrolysis degradation products coupled with the model compound oxidation results of Randall et al. (1938) and others. It is easy to criticize this model since it cannot account for such observed properties in aquatic humic substances as color; however, Christman et al. (1987) stress that this model is a concept which explains the degradation results and should only be utilized as a basis for formulating new degradation schemes and structural hypotheses. This statement should, in fact, hold true for other proposed humic structures.

Pitfalls in the degradative approach

The preceding discussion has served to illustrate the manner in which the degradative approach to humic structural analysis is most often applied. It should be emphasized that there are potential pitfalls inherent in the application of chemical degradation which all workers need to recognize. These pitfalls are of two general types.

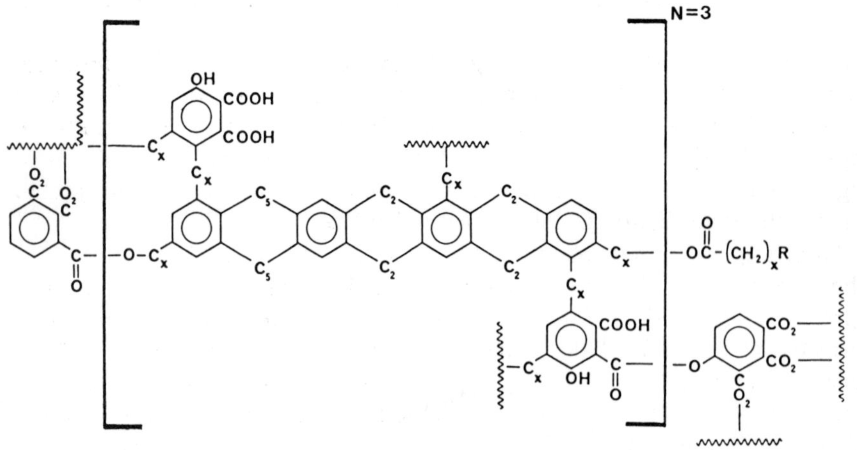

Fig. 3—"Aromatic matrix" model for aquatic humic macromolecular structure proposed by Christman et al. (1987).

Experimental pitfalls. These include such actions as selection of unoptimized reaction conditions, improper use of reaction blanks to detect contamination, selection of an extraction procedure which fails to adequately recover degradation products, improper identification of degradation products, and inaccurate degradation product yield measurements. Owing to the widespread use of GC/MS, the latter two warrant special discussion.

The term "identification" as applied to qualitative GC/MS analysis of components in complex mixtures has no standard definition. Further, it is often impossible to discern from the experimental description of a given study exactly what criteria were employed to "identify" a given degradation product. Since the advent of modern digital computers a component is often said to be identified if its electron impact (EI) mass spectrum resembles that of a library spectrum stored in the computer's memory, utilizing criteria sometimes known only to the computer. Other opinions have been advanced as to what should constitute an "identification" (Millington and Norwood 1986). It is essential that any worker employing the degradative approach clearly state the criteria utilized for the structural assignment of each degradation product so that others may independently judge the validity of structural inferences.

Accurate degradation product yield measurements require that additional points be taken into consideration. These include matrix effects and product losses during extraction and concentration procedures, variations in the behavior of the separation device (i.e., gas chromatograph), and variations in the detector (i.e., mass spectrometer). Reported degradation product yield measurements which are not computed utilizing known method recoveries for individual compounds, proper internal standards, and adequate instrument calibration and standardization procedures should be viewed with caution.

Interpretation pitfalls. These include the use of degradation products formed in low yield for major structural inferences and artifacts. The issue of "artifact" formation is a pervasive one in the application of the degradative approach. In this context, an "artifact" can be defined as an identified degradation product whose formation pathway is incorrectly interpreted leading the investigator to false structural inferences.

An example of a possible artifact-generating reaction was investigated by Cheshire et al. (1968). These workers detected 3,5-dihydroxybenzoic acid as a product from the KOH fusion at elevated temperature of synthetic polymers derived from *o*- and *p*-benzoquinones and furfural, as well as from a peat derived humic acid. Many degradation processes, such as alkaline CuO oxidation, also produce *m*-dihydroxy aromatic structures from various humic substances leading investigators to hypothesize the existence of such structures in the humic macromolecular structure. This is of importance to water chemists since *m*-dihydroxy aromatic structures react readily with aqueous chlorine to produce chloroform, a suspected carcinogen, in drinking water.

Based on their synthetic polymer degradation results, Cheshire et al. (1968) hypothesized that the 3,5-dihydroxybenzoic acid formed from the humic substance was a secondary product "resulting from the breakdown of the complex products into relatively simple units which recombine under the fusion conditions." This conclusion is debatable since the synthetic "polymers" were not well characterized and secondary reactions could have taken place during the "polymerization" to produce *m*-dihydroxy aromatic structures.

It may be impossible to demonstrate the absence of artifact formation from the chemical degradation of complex natural products like humic substances. However, there appears to be little hard scientific evidence supporting many of the claims for artifact formation that have been presented.

Structural inferences from degradative studies

Numerous structural inferences have been drawn based on the application of chemical degradation procedures to humic substances extracted from soils, sediments, peats, lakes and rivers, and the oceans. Chemical models for humic macromolecular structure, such as the two described earlier, have been proposed which incorporate these structural inferences, but no attempt will be made to summarize this literature. Instead, two important conclusions will be emphasized which seem important to this author from the wealth of chemical degradation data that exists in the scientific literature.

Isolated aromatic nuclei are important in the macromolecular structure of humic substances of terrestrial origin.   The majority of studies indicate that these aromatic nuclei are highly substituted with either crosslinking aliphatic side chains or functional groups such as carboxyl, hydroxyl, and methoxyl. For certain humic substances studied (especially those of aquatic origin) a substantial percentage of these nuclei appear not to possess an aromatic C-O function, i.e., they are not representative of intact lignin/phenolic structures. It is thus apparent that phenolic moieties have been overemphasized in hypothetical humic structural models to date. More emphasis should be placed on discerning the origin and formation mechanisms for other types of aromatic nuclei.

Aliphatic structures have greater importance in the macromolecular structure of many humic substances than previously believed.   This is especially true when one considers freshwater aquatic and marine humic substances. The majority of published structural models, including the "building block" model for soil humic structure and the "aromatic matrix" model for aquatic humic structure, emphasize the aromatic structural components. Far too little attention has been paid to the nature and origin of aliphatic carbon in humic structure. This historical oversight has led many investigators to potentially false and dangerous assumptions.

A good example is the hypothesis formulated in 1974 by J. J. Rook (Rook 1974), that phenolic nuclei present in the macromolecular structure of aquatic humic substances were responsible for the formation of trihalomethanes in drinking water during aqueous chlorination. This hypothesis was based on the existing body of chemical degradation data regarding soil and aquatic humic substances. Rook's ideas have led other investigators to conduct detailed studies of the reactions of various phenols with aqueous chlorine (Boyce and Hornig 1983, for example). However, the importance of aliphatic structures in aquatic humic substances has since been demonstrated, and Reckow and Singer (1985) have formulated a structural model for aquatic humic structure which rationalizes the formation of most of the major aqueous chlorination products from an aliphatic precursor structure. This controversy regarding the origin of organohalides in drinking water has left governmental regulators in a dilemma.

THE NONDEGRADATIVE APPROACH

Overview and examples

As stated previously, the nondegradative approach seeks to analyze isolated humic substances without chemical or physical alteration utilizing techniques which allow structural inferences to be drawn. The techniques utilized include UV-visible and infrared spectrophotometry, electron spin resonance spectrometry and NMR, X-ray analysis, electron microscopy and electron diffraction, viscosity and surface tension measurements, and various titrimetric methods. It has become apparent in recent years that the most promising nondegradative technique for humic structural analysis is NMR.

The utility of NMR for humic structural analysis is primarily due to two fundamental instrumental advances. The first of these is the Fourier transform technique described in detail by Fukushima and Roeder (1981). Basically, the technique allows the rapid acquisition and averaging of many individual spectra to produce an average spectrum with greatly enhanced signal-to-noise ratio and thus, sensitivity. This is critical for samples such as humic substances which have limited solubility in NMR solvents. It is also important for the study of relatively dilute spin systems, such as ^{13}C (Levy et al. 1980). The second major instrumental advance is the crosspolarization/magic-angle spinning (CP/MAS) technique which allows the acquisition of ^{13}C NMR spectra from solid samples, somewhat alleviating the sensitivity and other problems inherent in solution NMR experiments. A more detailed discussion of the CP/MAS ^{13}C NMR technique is beyond the scope of this chapter. The interested reader is advised to consult the monographs by Levy et al. (1980) and Fukushima and Roeder (1981).

To date, CP/MAS ^{13}C NMR spectra have been obtained from humic substances isolated by a variety of procedures from numerous matrices and environments. Hatcher and co-workers at the U.S. Geological Survey have published such spectra of aquatic and marine sedimentary humic substances, aquatic humic substances, soil- and peat-derived humic substances, and isolated aquatic humic substances (Hatcher et al. 1983, for example). Figure 4 shows three spectra acquired by this group including a humic acid derived from Everglades peat, a humic acid derived from a Histosol soil, and a

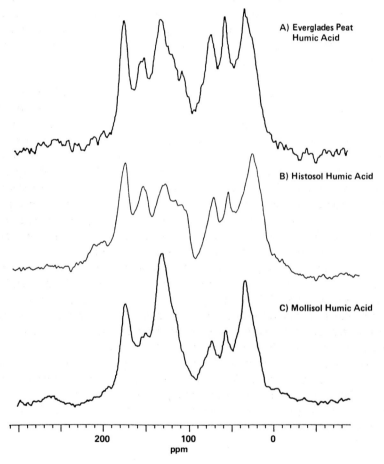

A) Everglades Peat Humic Acid

B) Histosol Humic Acid

C) Mollisol Humic Acid

200   100   0
ppm

Fig. 4—CP/MAS ^{13}C NMR spectra of three isolated humic acids.

humic acid derived from a Mollisol soil. The signals in these NMR spectra represent the different types of carbon atoms present in these materials. Further, CP/MAS spectra are generally taken to be quantitative, that is, the relative intensities of the various resonance signals are proportional to the relative amounts of the various carbon atoms in the material under investigation.

For example, Fig. 5(A) shows a spectrum (Norwood 1985) of an aquatic fulvic acid (Singletary Lake, North Carolina) which can be divided into four principal resonance regions. The relative intensities, as represented by peak areas integration, coupled with the positions of the resonance signals in each region allow structural inferences to be drawn. First, it is apparent that this humic substance is highly aliphatic in nature (regions I and II) as well as highly acidic (region IV). Aromatic carbon is also apparent (region III), but relatively little of this carbon appears to be substituted with heteroatoms such as oxygen. This implies that intact lignin and other phenolic units are only minor structural components of this material. Note, however, that this is not the case for certain soil-derived humic substances such as those shown in Fig. 4.

Further structural inference can be derived from the ^{1}H NMR spectrum (Norwood 1985) shown in Fig. 5(B). This spectrum was obtained from an NaOD/D_2O solution of the humic substance. Four spectral regions are again apparent which represent the types and relative amounts of all non-deuterium exchangeable protons in the aquatic fulvic acid. As in the CP/MAS spectrum, aliphatic structures are shown to be of significance. The lack of significant aromatic proton (Ar-H) coupled with the amount of aromatic carbon indicates that aromatic structures are highly substituted, most likely with aliphatic side chains. These data and structural inferences are supportive of those of Liao et al. (1982) described earlier in this paper. It is thus apparent that the degradative and nondegradative approaches are complementary.

Schnitzer and his co-workers have also employed numerous nondegradative techniques to examine soil-derived humic substances. Preston and Schnitzer (1984) acquired ^{13}C NMR spectra of several methylated soil humic and fulvic acids dissolved in deuterochloroform. They interpreted resonance signals in the aromatic carbon region of these spectra to be indicative of aromatic carbons substituted by methylated carboxyl groups. These signals were correlated with others in the aliphatic region due to added methyl groups on these carboxyls. They concluded that carboxyl groups in these humic substances are almost exclusively attached to aromatic rings and suggested that these data were support of the "building block" model for soil humic macromolecular structure.

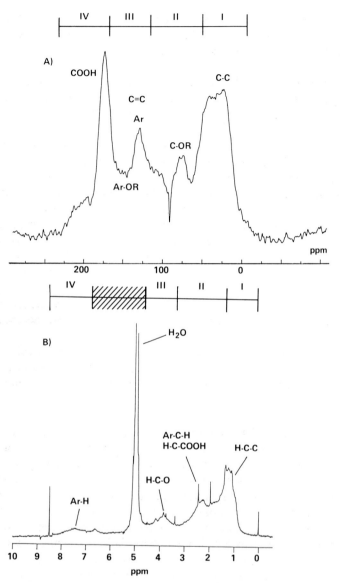

Fig. 5—CP/MAS (A) and ¹H (B) NMR spectra of Singletary Lake fulvic acid.

Pitfalls in the nondegradative approach

It is clear that the potential for combining degradative and nondegradative approaches to attack specific structural problems is substantial. There are few examples in the scientific literature of such a combination.

Nondegradative techniques have their potential pitfalls as do degradative techniques and it is essential that the nature of these pitfalls be recognized. As with the degradative approach, there are two general types.

Experimental pitfalls. Each particular nondegradative technique has its own unique experimental pitfalls. This discussion will utilize NMR as an example since it represents one of the most promising nondegradative analytical techniques.

The scientific literature is filled with studies involving the acquisition of CP/MAS ^{13}C NMR spectra of various humic substances. Without fail, investigators utilize these spectra to derive the quantitative carbon distributions of their particular sample, usually including in their discussions a few "hand-waving" statements regarding the quantitative nature of the CP/MAS technique. Few workers, however, bother to ensure that their NMR acquisition parameters have been optimized for quantitative analysis, even though the experiments to do so have been described (Wilson et al. 1983) and are not difficult to accomplish. Further, the non-quantitative nature of ^{13}C NMR spectra obtained from dissolved samples has long been recognized (Levy et al. 1980). Workers employing solution techniques must pay particular attention to acquisition parameters when interpreting their spectra quantitatively.

Any ^{13}C NMR spectrum, whether CP/MAS or solution, which is to be interpreted quantitatively should include evidence that all acquisition parameters have been optimized for such interpretation. Those that do not should be viewed with caution.

Interpretation pitfalls. The theoretical assignment of NMR chemical shift to various types of ^{13}C and ^{1}H nuclei is in an early stage of development (Levy et al. 1980). Spectra are interpreted, therefore, by correlating the various resonance signals with those of known chemical structures. In complex macromolecular systems such as humic substances, nuclei can experience a wide variety of chemical environments producing a wide variety of chemical shifts. These may not correlate well with chemical shifts of simple compounds and care must be taken to avoid overinterpretation. Chemical shift assignments that appear too specific and are not accompanied by confirming data should be viewed cautiously, especially if they support the investigator's preconceived notions.

Structural inferences from nondegradative studies

It should now be obvious that the information obtained from the degradative and nondegradative approaches to humic structural analysis is complementary. Therefore, two overall conclusions will be stated, the first of which has generally not been supported by degradative studies and the second of which has.

Humic substances isolated from different environments and difference matrices differ greatly in chemical structure. CP/MAS ^{13}C NMR experiments clearly demonstrate dramatic structural differences in humic substances isolated from marine, terrestrial aquatic, soil, and other matrices. Hatcher et al. (1981) have shown that even certain isolated soil humic substances can differ widely in carbon distribution while others appear quite similar. These differences usually involve the relative amounts of aliphatic versus aromatic carbon. Degradative studies have not emphasized these differences to the same extent as nondegradative studies because of the low degradation product yields often observed and the vigorous nature of many degradation reactions which tend to destroy aliphatic structures.

Aliphatic structures have greater importance in the macromolecular structure of many humic substances than previously believed. This conclusion is worth restating since most studies, both degradative and nondegradative, have tended to concentrate on the aromatic features of humic structure. In many humic substances aliphatic structures are dominant. This is especially true for freshwater aquatic and marine humic substances. It is clearly time for investigators employing degradative approaches to consider ways of understanding this aliphatic structural component.

CONCLUSIONS

In this author's opinion we are on the verge of major advances in the understanding of the nature and chemical structure of humic substances. These advances will be made possible by the continuing rapid progress being made in the development of analytical techniques. It is appropriate at this time to mention a few of these techniques along with their potential benefits:

1) combined liquid chromatography/mass spectrometry (LC/MS) and fast atom bombardment mass spectrometry (FAB-MS) for the analysis of degradation products not amenable to GC/MS (FAB-MS holds the yet to be realized promise of the direct analysis of humic substances by mass spectrometry);
2) high field, CP/MAS ^{13}C NMR to provide increased spectral resolution;
3) solid-state ^{1}H NMR to better complement CP/MAS spectra;
4) alternate CP/MAS pulse techniques such as dipolar dephasing (Wilson et al. 1983) which are designed to address specific carbon atom types in the humic structure. With such techniques to aid both degradative and nondegradative approaches, it behooves all investigators to realize that

these approaches are indeed complementary and to proceed forward accordingly.

REFERENCES

Boyce, S.D., and Hornig, J.F. 1983. Reaction pathways of trihalomethane formation from the halogenation of dihydroxyl model compounds for humic acid. *Env. Sci. Tech.* **17**: 202–211.

Cheshire, M.V.; Cranwell, P.A.; and Haworth, R.D. 1968. Humic acid–III. *Tetrahedron* **24**: 5155–5167.

Christman, R.F.; Norwood, D.L.; Seo, Y.; and Frimmel, F.H. 1987. Oxidative degradation of humic substances from freshwater environments. In: Humic Substances II: In Search of Structure. New York: John Wiley and Sons, in press.

Dienert, F. 1910. Research on fluorescent substances in the control of disinfection of water *C.R. Acad. Sci.* **150**: 487.

Fukushima, E., and Roeder, S.B.W. 1981. Experimental Pulse NMR. A Nuts and Bolts Approach. Reading, Massachusetts: Addison-Wesley Publishing Co.

Hatcher, P.G.; Breyer, I.A.; Dennis, L.W.; and Maciel, G.E. 1983. Solid-state ^{13}C-NMR of sedimentary humic substances: new revelations on their chemical composition. In: Aquatic and Terrestrial Humic Materials, eds. R.F. Christman and E.T. Gjessing, pp. 37-81. Ann Arbor, MI: Ann Arbor Science Publishers.

Hatcher, P.G.; Schnitzer, M.; Dennis, L.W.; Maciel, G.E. 1981. Aromaticity of humic substances in soils. *Soil Sci. Soc. Am. J.* **45**: 1089–1094.

Hayes, M.H.B., and Swift, R.S. 1978. The chemistry of soil organic colloids. In: The Chemistry of Soil Constituents, eds. D.J. Greenland and M.H.B. Hayes, pp. 179–320. New York: Wiley-Interscience.

Levy, G.C.; Lichter, R.L.; and Nelson, G.L. 1980. Carbon-13 Nuclear Magnetic Resonance Spectroscopy, 2nd ed. New York: Wiley-Interscience.

Liao, W.T.; Christman, R.F.; Johnson, J.D.; and Millington, D.S. 1982. Structural characterization of aquatic humic material. *Env. Sci. Tech.* **16**: 403–410.

Millington, D.S., and Norwood, D.L. 1986. Application of combined gas chromatography/mass spectrometry to the identification and quantitative analysis of trace organic contaminants. In: Organic Carcinogens in Drinking Water, eds. N.M. Ram, E. Calabrese, and R.F. Christman, pp. 131–152. New York: John Wiley and Sons.

Norwood, D.L. 1985. Aqueous Halogenation of Aquatic Humic Material: A Structural Study. Dissertation: Univ. North Carolina, Chapel Hill.

Preston, C.M., and Schnitzer, M. 1984. Effects of chemical modifications and extractants on the carbon-13 NMR spectra of humic materials. *Soil Sci. Soc. Am. J.* **48**: 305–311.

Randall, R.B.; Benger, M.; and Croocock, C.M. 1938. The alkaline permanganate oxidation of organic substances selected for their bearing upon the chemical constitution of coal. *Proc. Roy. Soc. Lon. A.* **165**: 432–452.

Reckow, D.A., and Singer, P.C. 1985. Mechanisms of organic halide formation during fulvic acid chlorination and implications with respect to preozonation. In: Water Chlorination: Chemistry, Environmental Impact and Health Effects, eds. R.L. Jolley, R.J. Bull, W.P. Davis, S.Katz, M.H. Roberts, and V.A. Jacobs, vol. 5, pp. 1229–1258. Chelsea, MI: Lewis Publishers.

Rook, J.J. 1974. Formation of haloforms during chlorination of natural waters. *Water Treat. Exam.* **23**: 234–243.

Schnitzer, M. 1978. Humic substances: chemistry and reactions. In: Soil Organic Matter, eds. M. Schnitzer and S.U. Khan, pp. 1–64. Amsterdam: Elsevier.

Schnitzer, M., and Khan, S.U. 1972. Humic Substances in the Environment. New York: Marcel Dekker.

Thurman, E.M. 1985. Organic Geochemistry of Natural Waters. Dordrecht, Netherlands: Martinus Nijhoff/Dr. W. Junk Publishers.

Wilson, M.A.; Pugmire, R.J.; and Grant, D.M. 1983. Nuclear magnetic resonance spectroscopy of soils and related materials. Relaxation of ^{13}C nuclei in cross polarization nuclear magnetic resonance experiments. *Org. Geochem.* **5**: 121–129.

Standing, left to right:
Horst Nimz, Michael Hayes, Wolfgang Ziechmann, Simon Visser

Seated (center), left to right:
Morris Schnitzer, Mike Perdue, Gudrun Abbt-Braun, Michael Bracewell

Seated (front), left to right:
Dan Norwood, Jan de Leeuw, Mick Wilson

Humic Substances and Their Role in the Environment
eds. F.H. Frimmel and R.F. Christman, pp. 151–164
John Wiley & Sons Limited
© S. Bernhard, Dahlem Konferenzen, 1988.

The Characterization and Validity of Structural Hypotheses
Group Report

J.M. Bracewell, Rapporteur
G. Abbt-Braun   E.M. Perdue (Moderator)
J.W. de Leeuw   M. Schnitzer
M.H.B. Hayes   S.A. Visser
H.H. Nimz   M.A. Wilson
D.L. Norwood   W. Ziechmann

INTRODUCTION

In biochemical investigations we are accustomed to expect an ultimate point to be reached at which a defined chemical structure can be presented which is able not only to rationalize and predict the observed chemical behavior but is itself uniquely defined. This is so, for example, with established polypeptides and polysaccharides. In other cases no uniquely defined structure may be determined because of a random element in the way the building blocks are linked. This occurs with lignin, for example, although the phenylpropanoid monomers are fully characterized and the several modes of linkage understood. Lignin chemists are accustomed to express this knowledge in terms of a structural representation which, though it may not represent any real unique molecule, illustrates the monomers and types of linkage.

Humic materials have proved still more difficult to investigate and characterize, but researchers have nevertheless attempted to express the results of their investigations by structural representations. These are admittedly hypothetical but are an attempt to rationalize chemical behavior as far as possible. Such representations are sometimes referred to as structural "models," though this term should not be confused with that of model

151

compounds as used in experiments to simulate the genesis of humic substances (see Ertel et al., this volume).

The use of these purely hypothetical structures has frequently led to misunderstandings in the literature and at meetings. In this report, therefore, we attempt to look at what purposes they serve, examine how they are built up, how they are being modified in the light of modern analytical methods, and in what respects they represent an average or composite of what observation tells us.

FUNCTIONS SERVED BY HYPOTHETICAL STRUCTURAL MODELS

The main uses of hypothetical structural models are (a) as a way to represent the average properties of a complex mixture, such as humic substances, and (b) to help the formulation of new hypotheses regarding humic structure and the development of new experimental schemes for their investigation.

The origin and development of structural models

Structural models first arose from a need to represent the known chemistry of humic substances in terms of structures which, though hypothetical, are based on chemically perceived experience. Formerly this chemical knowledge stemmed from various degradative procedures, spectroscopic data, and pH solvation behavior. The structural representations which subsequently evolved from this foundation should be regularly updated in the light of new evidence and techniques, as can be seen, for comparison, with the case of lignin studies. For example, the nonrealization of the importance of aliphatic macromolecules caused their exclusion from early structural models. Later results from ^{13}C NMR and analytical pyrolysis have emphasized their significance, so an attempt must be made to build them in.

In order to evolve model hypotheses by the combination of information, a recommended procedure would be to start from the average elemental composition, functional and structural group analyses and molecular weight. Taking the structural groups one tries to work towards subfunctional group structure, e.g., whether a carbonyl is adjacent to a hydroxyl or methyl. In this process there is a duty to be exact within the constraints of knowledge as regards known properties and interactions. What remains as the unknown part of the structure after this process is carried out will hopefully become ever less and less. At each stage, however, one may intelligently hypothesize about it within the constraints of composition and chemical nature. On this point, however, differences of opinion arise and are considered below.

One danger is that prior assumptions may be incorporated in addition to what can strictly be deduced from experiment, and these features, though experimentally ungrounded, tend to be propagated by others to future structural models. Another danger is the natural tendency to focus on those aspects of the real structures which are thought to be known, whereas new research should continually seek the aspects which are unknown.

Limitations and misinterpretations of composite structures

It is clearly difficult to represent even approximately, by a single scheme, an entity which ranges in molecular weight between 10^6 and less than 10^3 daltons, and which very likely has no permanent or regular structure and may or may not have characteristic fundamental structural units or a stable core. Structural representations therefore should preferably refer to specific fractions and we may need functional models for use within a more narrowly defined context. For example, we may construct a representation at the level of primary structure from structural groups or at a tertiary level from the knowledge of random coil formation under changing pH conditions.

It is difficult to incorporate the quantitatively small features, and misinterpretations in the use of hypothetical molecular models can arise from the typically qualitative nature of such models, which tends to overemphasize certain less abundant features. Molecular models should be made as quantitative as possible within the constraints of the experimental evidence.

Further misinterpretation arises from a wrong perception and use of models which are only intended as working structural hypotheses. An example is given by Norwood (this volume) where the adoption of an early model for soil humic materials when dealing with aquatic materials led to a diversion of much research effort during the investigation of organohalides formed during water chlorination. This occurred because phenolic groups, featured in these earlier structures, were thought to be responsible, whereas other types of structure are now known to be far more significant for aquatic materials.

Differences of approach to modelling

It is clear that some scientists value structural models as an expression or summation of the results of investigations for the purposes of comparison, education, and the expression or prediction of properties. Others, however, are adamant that models are not justified by the small size of fragments we have identified at present, and that one should only express what has been rigorously tried and tested. This to some extent reflects a difference of scientific approach between inductive and strictly deductive schools of thought. Good science is usually recognized as a process which incorporates both activities.

THE EFFECT OF ISOLATION/FRACTIONATION ON OUR VIEW OF HUMIC SUBSTANCES

The avoidance of artifact formation in our schemes of isolation and fractionation must carry a high priority if we are adequately to interpret structural data. Samples are extracted from the solid phase (soils and sediments) by quite different methods than from the dissolved phase (aquatic humic materials), and the two methods are therefore considered separately.

Humic materials extracted from the solid phase

We strongly advise the removal of certain entities which may appear in the later stages of a fractionation and cloud our interpretations of the essential nature of the fractions. These are chiefly the nonpolar and the biologically undegraded matter.

Biologically undegraded material will clearly distort the properties of any fraction into which it is incorporated. Furthermore, when extracted with alkali, it gives products which simulate those of humic substances (Ertel and Hedges 1985). Its removal should therefore be effected by a densimetric separation such as flotation or density centrifugation (Dyrkacz and Horwitz 1982; Dyrkacz et al. 1984).

The nonpolar fraction should also be removed as far as possible. Organic solvent extractions remove only ca. 3–6% whereas recent evidence shows that supercritical ethanol/H_2O extraction at elevated temperatures (Schnitzer and Preston 1987) can remove on the order of 50%. Clearly this is of major importance and requires further investigation, preferably by employing a fluid which can be used at room temperature such as CO_2. The extract itself is evidently of considerable interest and requires further characterization.

The various isolation techniques depend on the solvation properties of humic substances in different solvents. On extraction with mild aqueous pyrophosphate at pH 7, the divalent and polyvalent metals, which are complexed with and render insoluble the humic macromolecules, form alternate complexes with the pyrophosphate. This allows the humic substances to solvate in water, and the process is further helped by the dissociation of acid groups, primarily carboxyls, at pH 7. As the pH is raised above 7, less strong acid groups such as phenols can dissociate, and so additional dissociable groups are ionized and solvate in the water. The acid groups are reformed as carboxyl, phenol, etc., when the pH is lowered, and precipitation variously takes place as the pH is lowered to 1. The relatively highly polar fulvic acids, however, stay in solution at pH 1.

The solvation mechanisms in organic solvents differ from those in water, and Hayes (1985) has summarized various properties of organic solvents which determine their suitabilities for use in the dissolution of humic substances. Special attention is drawn to the use of dimethylsulphoxide (DMSO) and of other dipolar aprotic organic solvents, with emphasis on the breaking of the hydrogen bonds which hold together the humic molecules in the H^+-exchanged form.

The hydrogen bonds are also broken when the acid groups ionize and the molecules are repelled by coulombic repulsion forces. The alkaline conditions used to achieve this, however, give rise to oxidation of the humic substances, and though this can be partially suppressed by using an inert atmosphere (N_2), there is always the possibility of oxidation from OH^-.

The lower molecular weight humic acids are extracted in the neutral metal complexing media (such as pyrophosphate), and the highest molecular weight materials are extracted in heated alkaline (NaOH) solutions, as shown for carefully fractionated samples by Cameron et al. (1972). Similar data are not available for extracts in acidified DMSO, but it is clear that the humic substances isolated in this solvent have less of the stronger acidic groups and more of the weaker ones (phenolic groups) than have the NaOH extracts.

Humic materials extracted from the dissolved phase

Isolation and separation in these cases is brought about in more dilute solutions on macroreticular resins. In the presence of acid and alkaline media there is a risk of ester linkage hydrolysis which could seriously affect carboxyl content and hence metal binding and titratable acidity, for example, in investigations of the role of humic substances in acid rain studies. There is a risk also of hydrolysis and of oxidation when alkaline solutions are used to desorb the humic substances from the sorbing media. Contact with ammonia during preparation can profoundly affect binding and create artifacts. The use of DEAE (abbreviations, see Appendix 2) celluloses is being proposed, as these do not require the use of pH values removed from neutrality. A critical review with recommendations is given by Aiken (1985).

On the positive side, methods of resin extraction are appearing which are increasingly selective for fractions rich in a particular structural grouping, such as aliphatic chains, carbohydrates, carboxyl, and phenolic hydroxyl (Thurman, personal communication). For example, an aquatic fulvic acid can be separated on a weak base-exchange resin such that the carboxyl is ionized but the phenolic hydroxyl is not, leading to fractions enriched in each grouping. Another example is the isolation of nitrogen-rich humic materials on cation exchange resins (Schnitzer and Khan 1972).

EVIDENCE FOR STRUCTURE FOLLOWING DEGRADATIVE TREATMENTS

In any degradative treatment an inverse relationship is evident between selectivity and yield. A high yield of small fragments may be obtained which, though they may be firmly identified, bear little structural information. Conversely a very mild treatment may give a lower yield of fragments which are large and bear much structural detail but are difficult to identify. This dilemma is apparent in both chemical and thermal degradative methods.

Chemical degradative methods

The manner in which evidence may be combined to give a consistent structural picture is exemplified by much of the work with permanganate and alkaline cupric oxide degradations, though other oxidants such as hydrogen peroxide, perchloric acid, and nitric acid have been used. In these studies all possible benzene carboxylic acids were identified as products. Though these could have arisen from aliphatic side chain oxidation or fused ring degradation, they were also obtained from alkaline cupric oxide treatment, which does not degrade fused aromatic rings. Some products showed a $-CH_2-COOH$ substituent indicating that a side chain was attached to some of the rings at least. Aliphatic dicarboxylic acids, $HOOC-(CH_2)_n-COOH$ ($n=0$ to 8), were found, which could arise from unsaturated aliphatic chain structures as follows

$$R-CH=CH-(CH_2)_n-CH=CH-R' \rightarrow R-COOH + HOOC-(CH_2)_n-COOH + HOOC-R'.$$

A picture, therefore, emerges of aromatic nuclei, highly substituted with crosslinking aliphatic side chains or with functional groups such as carboxyl.

In reductive methods, zinc dust distillation gave aromatic fused ring compounds, leading initially to the idea of a stable core. Sodium amalgam reduction gave di- and trihydroxybenzenes, also with $-COOH$ or $-CH_3$ substitutents, indicative of possible lignin origin. There was never more than one $-COOH$ in this case on any one ring, but clearly this could not have arisen as an artifact in a reductive system. Degradation with phenol cleaves the aliphatic linkages joining aromatic groups, suggesting structures such as $Ar-(CH_2)_n-Ar$ where Ar is an aromatic ring.

In such ways useful data emerge to give integrated evidence of structural units. The models which have arisen from this data are discussed by Norwood (this volume) and a review of the chemical degradative methods is given by Hayes and Swift (1978). It should be remembered that yields are of the order of 20%. We believe that much of this work such as permanganate oxidation should now be repeated using milder forms of treatment and at the same time following the detailed changes, not only in the extract but also in the residue, which is now possible using ^{13}C NMR, analytical pyrolysis, etc.

Thermal degradative methods

The selective rupture of the weaker bonds in a polymer structure can readily be brought about by analytical pyrolysis (Irwin 1982). Synthetic polymers and biopolymers are split into fragments closely related to the effective monomer unit (see section below). Fragments tend to be small where there are transfer problems as in pyrolysis-GC-MS or pyrolysis-EIMS, when yields are of the order of 10-15%. The information given is essentially, therefore, the molecular configuration of unit moieties within the larger structure and, as such, is essentially qualitative and, in the case of humic substances, complex (Saiz-Jimenez and de Leeuw 1986). This complexity can, however, be readily analyzed in terms of significant variations between types, for example soil humus types (Bracewell and Robertson 1984).

The pyrolysis method may be extended by the use of temperature programming combined with direct evaporation into a field ionization mass spectrometer source operating with high resolution. The "soft" ionization then gives unit fragments of molecular weight up to 1000 daltons or higher, showing much more potential for structural determination (Schulten 1984).

The possibility of a structural "core"

The concept of a "core" to humic structures in soils arose through the classical degradative experiments because of the inert residue obtained after removal of the biopolymers. Such residues are now amenable to modern analytical methods, such as NMR and analytical pyrolysis, and are currently receiving attention.

Perhaps the most helpful picture of such material is obtained by considering the humic substances not as thermodynamically stable, but as a kinetically cycling pool of organic moieties in the biosphere. Some components degrade and cycle rapidly, such as most carbohydrates and proteins, whilst others do so more slowly and result in greater accumulations. These include, for example, the aliphatic non-ester components of plant cuticles and other as yet unknown plant biopolymers (Nip et al. 1986). One way of obtaining this relatively inert material would be to seek out natural environmental situations where the process of early diagenesis has essentially removed the more labile biopolymers.

The possibility now opens up through NMR to operate with such materials not only by degradation but additively, for example by the placement of extra methyl groups and the observation of their molecular motion through relaxation times, giving information about the steric chemical environment.

Tertiary structure and the shape of humic macromolecules

Concepts derived from physicochemical data can be used to indicate the shapes which humic macromolecules can adopt. The fractionation of humic acids using procedures which employ gel chromatography and graded porosity membranes give fractions of relatively homogeneous molecular size (Cameron et al. 1972). Among these, the relationship between frictional ratio values and molecular weight, both obtained from ultracentrifugation, indicate molecules of the same general shape but increasing molecular size.

This data is best interpreted in a general way in terms of a random coil confirmation for humic substances in solution, though small molecules tend to be more linear. On acidification the charge on ionized groups is neutralized, and inter- and intramolecular hydrogen bonding causes shrinkage with exclusion of solvent from the matrix, leading to precipitation. Polyvalent cations have a similar effect, drawing the structure together by bridging two or more negatively charged groups. The interaction with metal or other organic substrates thus depends on their ability to penetrate the random coil matrix, a process which is to a considerable extent regulated by the cations neutralizing the charges and by the water content of the medium. For a detailed discussion see Hayes and Swift (1978).

MOLECULAR LEVEL ANALYTICAL TOOLS IN INVESTIGATIONS OF HUMIC SUBSTANCES AND BIOPOLYMERS

Modern analytical techniques, particularly spectroscopic techniques, are revealing a wide range of information on the structure and reactivity of humic substances. Two questions come into consideration, namely what types of information can be obtained and how quantitative it is.

The major techniques involved are mass spectrometry (MS), analytical pyrolysis, nuclear magnetic resonance (both solution and solid-state), Fourier-transform infrared spectroscopy, ultraviolet/visible absorption and fluorescence spectroscopy, and chromatographic methods such as gel permeation chromatography and high performance liquid chromatography (HPLC) with different phases. Insights into these methods and their relevance can be gained from studies of appropriate biopolymers, for example lignin, which are also useful since they are mostly precursors to humic substances. We should not forget computational procedures based on elemental and functional group analysis which can be used to place quantitative limits on possible structural composition (Perdue 1984). The first three of the above list are considered here as principal examples. An important consideration with them all is the degree of expertise required, not only in their operation but in their adequate interpretation.

Nuclear magnetic resonance (NMR)

The success of NMR arises in part because the final analysis appears simple, i.e., the assessment of structural group content from integrated areas of group-specific bands in a spectral plot. However, without a more detailed knowledge of NMR there are pitfalls that must be avoided. The humic chemist needs to know enough NMR theory to appreciate relaxation mechanisms even though for most samples the data obtained will be at least qualitatively reliable. Another limitation is sensitivity; about 300 mg of sample are usually needed for an adequate spectrum in 1 hour (solid) or 24 hours (solution), but as little as 10 mg if the sample contains over 50% C. At low C content ($<5\%$), solid samples such as soils may need removal of Fe^{3+}. Obtaining spectra from larger samples is limited by rotor design and coil size and frequently it is necessary to increase time of analysis at the expense of quantitation (Wilson 1987).

Spectral editing techniques based on pulse sequences may allow specific structural groups, for example nonprotonated aromatic carbons, to feature specifically in spectra and this possibility may give promising advances. Nevertheless, better sample fractionation would greatly increase the quality of information obtained by decreasing the overlap of bands. Derivatization with ^{19}F, ^{29}Si, ^{31}P, and ^{13}C reagents would also improve structural evidence (Wilson 1987).

Analytical pyrolysis

This technique combines degradation (thermal, including pulse and temperature programmed methods), separation (by gas chromatography (GC) or MS), and identification (by MS with various ionization techniques: electron impact (EI), chemical ionization (CI), field ionization (FI), field desorption (FD), fast atom bombardment (FAB) or high resolution MS, or MS/MS). The temperature–time profile is required to be highly reproducible and the products formed must be rapidly removed from the pyrolysis zone. Sample sizes are 10–50 µg, requiring highly homogenized samples. The complex pattern of products may be interpreted by reference to those of known biopolymers or may be used to investigate significant differences of sample type using multivariate statistical analyses. A review of the techniques is given by Irwin (1982) and of their applications to humic substances by Bracewell et al. (1987).

The full potential of these methods has probably not yet been realized for in its more advanced and potentially useful forms, such as pyrolysis-FIMS (Haider and Schulten 1985; Schulten et al. 1987), the amount of data generated cannot at present be addressed without reduction at the expense of relevant information. Thus one pyrolysis may generate 1000 mass ions, each of which may be split by high resolution into 5 or 6 molecular ions, of which any one may be investigated for identification by MS/MS. Selection of ions with respect to origin, variability, and correlation with specific properties of humic substances, using statistical pattern recognition methods, may be a possible advance on this problem and enable us to recognize the relevant substructural moieties which reflect the structure of the original macromolecules.

Mass spectrometry

New developments in MS establish it as an instrument which, in its own right, combines degradation, separation, and identification in a single step within the ion source. Advanced soft ionization methods such as FI, FD, and FAB (especially suitable for polar materials) can produce mass fragments up to several thousand daltons. High field magnets coupled with extended geometry now enable mass values of this magnitude to be analyzed (Schulten and Lattimer 1984). The combination of liquid chromatography (LC) with MS also opens up the possibility of separation of polar materials in solution phase followed by intermediate MS identification.

As with the chemical degradative methods, we here emphasize the complementary nature of these and other techniques and recommend their simultaneous application in cooperative studies to known biopolymers (e.g., Saiz-Jimenez et al. 1987) and also to the now available reference humic substances provided by the International Humic Substances Society (see Thurman et al., this volume, for details).

THE STRUCTURAL BASIS OF AVERAGE PROPERTIES FROM DIFFERENT METHODS OF CHARACTERIZATION

Our expectation is that virtually all observed properties of a humic substance are the result of a continuum of sometimes widely ranging contributions. It is important, therefore, to take this variation into account when dealing with structural hypotheses and their validity, for example when considering "average" structural features such as the average degree of aromatic substitution obtained by a combination of ^1H and ^{13}C NMR band intensities and elemental composition. As another example, the degree of polydispersity is sufficient to give a wide difference between number average and weight average molecular weights.

The nature of averaged properties

Any measured average property P of a composite, such as a humic substance, can be expressed as a statistically weighted average of individual component properties P_i (Appendix 1, eq. 1). The value, within limits, of such properties is exemplified by the carbon content C, which will remain a useful intensive constant as long as experimental manipulations do not alter the relative masses m_i of the components (Appendix 1, eq. 2). Such averaged properties cannot, however, be converted into others in a similar manner to the behavior of a pure component. For example, a measured average equivalent spherical radius $\bar{r}$ cannot be converted to an average molar volume $\bar{V}$ by taking $(4\pi/3)\bar{r}^3$ (Appendix 1, eqs. 3).

In considering the chemical reactivity of humic substances, we have to deal with a distribution of sites with varying reactivity, and this will profoundly influence the type of interactions and their reaction kinetics. Consider, for example, the reaction with excess chlorine at constant pH, producing chloroform. A pseudo-first-order rate "constant" k_{obs} can be defined to express the proportionality between rate of chloroform production and total dissolved organic carbon (DOC) (Appendix 1, eq. 4.1). However, this overall rate represents a summation of individual component rates (eq. 4.2) and clearly the most reactive sites will be exhausted during the early stages of reaction, leading to a continuous decrease of k_{obs} with time. This decrease is sometimes represented by two straight line gradients for practical prediction purposes, which amounts to a gross oversimplification into a two-site model. The representation of properties such as reactivity in terms of statistical distributions rather than averages, offers therefore a type of structural model which accounts more closely for the observed behavior. Examples of the application of such distribution models to interactions with acid functional groups are given by Perdue (1985).

Statistics at molecular level

It is now becoming possible to examine individual components at the molecular level. A mass spectrometer operating in soft ionization mode effectively gives an actual distribution of molecular masses generated in the degradation process (thermal or otherwise). This information, as previously noted, can unfortunately be too large and unwieldy for practical operations without some data reduction. Nevertheless such distributions have been used practically, with the aid of statistical pattern recognition methods, to predict actual properties and to indicate the variation of structural components (Bracewell et al. 1987; Windig and Meuzelaar 1984).

With more molecular level information becoming available, certain structural patterns will hopefully begin to emerge in a statistical sense if we adopt such a pattern recognition approach.

SUMMARY OF THE GROUP'S RECOMMENDATIONS FOR FURTHER RESEARCH

1) The use of supercritical fluids to pre-extract humic substances for nonpolar

materials should be further investigated from the viewpoint of structural studies. Problems with the ionizable groups will, however, need to be addressed.

2) A search should be made for natural environmental situations where early diagenetic alteration has essentially removed the biopolymers, enabling examination of the relatively biostable residue or "core" by modern instrumental methods such as NMR.

3) In addition to the subtractive degradation of humic substances, the addition of labels should be tried. Studies of the behavior of these labels, e.g., in NMR experiments, should then lead to insights into the immediate environment of the label and hence the reactive site in the humic substance.

4) Soft-ionization high resolution MS methods should be applied to enable larger fragments of humic structure to be examined.

5) Oxidative and reductive degradations of humic substances should be repeated using milder forms of earlier treatments, and following the changes occurring in detail, especially in the residue, using ^{13}C NMR, analytical pyrolysis, etc. This should be tried also with model compounds such as those containing aliphatic chains.

6) Statistical trends in molecular level information should be examined over groups of humic samples with the aim of identifying characteristic features of humic materials and their specific properties.

REFERENCES

Aiken, G.R. 1985. Isolation and concentration techniques for aquatic humic substances. In: Humic Substances in Soil, Sediment and Water, eds. G.R. Aiken, D.M. McKnight, R.L. Wershaw, and P. MacCarthy, pp. 363–385. New York: Wiley.

Bracewell, J.M.; Haider, K.; Larter, S.R.; and Schulten, H.-R. 1987. Thermal degradation relevant to structural studies of humic substances. In: Humic Substances II: In Search of Structure (IHSS), eds. M.H.B. Hayes, R.L. Malcolm, and R.S. Swift. Chichester: Wiley, in press.

Bracewell, J.M., and Robertson, G.W. 1984. Characteristics of soil organic matter in temperate soils by Curie point pyrolysis-mass spectrometry. J. Soil Sci. 35: 549–558.

Cameron, R.S.; Thornton, B.K.; Swift, R.S.; and Posner, A.M. 1972. Molecular weight and shape of humic acid from sedimentation and diffusion measurements on fractionated extracts. J. Soil Sci. 23: 394–408.

Dyrcacz, G.R.; Bloomquist, C.A.A.; and Ruscic, L. 1984. High resolution density variations in coal separations by density gradient centrifugation. Fuel 63: 1166–1174.

Dyrcacz, G.R., and Horwitz, E.P. 1982. Separation of coal macerals. Fuel 61: 3–12.

Ertel, J.R., and Hedges, J.I. 1985. Sources of sedimentary humic substances: vascular plant debris. Geochim. Cosmochim. Acta 49: 2097–2107.

Haider, K., and Schulten, H.-R. 1985. Pyrolysis-field ionization mass spectrometry of lignins, soil humic compounds and whole soil. J. Anal. Appl. Pyrol. 8: 317–331.

Hayes, M.H.B. 1985. Extraction of humic substances from soil. In: Humic Substances in Soil, Sediment and Water, eds. G.R. Aiken, D.M. McKnight, R.L. Wershaw, and P. MacCarthy, pp. 329–362. Chichester: Wiley.

Hayes, M.H.B., and Swift, R.S. 1978. The chemistry of soil organic colloids. In: The Chemistry of Soil Constituents, eds. D.J. Greenland and M.H.B. Hayes, pp. 179–320. Chichester: Wiley.

Irwin, W.J. 1982. Analytical Pyrolysis: A Comprehensive Guide. New York: Marcel Dekker.

Nip, M.; Tegelaar, E.W.; Brinkhuis, H.; de Leeuw, J.W.; Schenck, P.A.; and Holloway, P.J. 1986. Analysis of modern and fossil plant cuticles by Curie-point Py–GC and Curie-point Py–GC–MS: recognition of a new, highly aliphatic and resistant biopolymer. *Org. Geochem.* **10**: 769–778.

Perdue, E.M. 1984. Analytical constraints on the structural features of humic substances. *Geochim. Cosmochim. Acta* **48**: 1435–1442.

Perdue, E.M. 1985. Acid functional groups of humic substances. In: Humic Substances in Soil, Sediment and Water, eds. G.R. Aiken, D.M. McKnight, R.L. Wershaw, and P. MacCarthy, pp. 493–526. New York: Wiley.

Saiz-Jimenez, C.; Boon, J.J.; Hedges, J.I.; Hessels, J.K.C.; and de Leeuw, J.W. 1987. Chemical characterization of recent and buried woods by analytical pyrolysis: comparison of pyrolysis data with ^{13}C NMR and wet chemical data. *J. Anal. Appl. Pyrol.*, in press.

Saiz-Jimenez, C., and de Leeuw, J.W. 1986. Chemical characterization of soil organic matter fractions by analytical pyrolysis–gas chromatography–mass spectrometry. *J. Anal. Appl. Pyrol.* **9**: 99–119.

Schnitzer, M., and Khan, S.U. 1972. Humic Substances in the Environment. New York: Marcel Dekker.

Schnitzer, M., and Preston, C.M. 1987. Supercritical gas extraction of a soil with solvents of increasing polarities. *Soil Sci. Soc. Am. J.* **51**: 639–646.

Schulten, H.-R. 1984. Relevance of analytical pyrolysis to biomass conversion. *J. Anal. Appl. Pyrol.* **6**: 251–272.

Schulten, H.-R.; Abbt-Braun, G.; and Frimmel, F.H. 1987. Time-resolved pyrolysis field ionization mass spectrometry of humic material isolated from freshwater. *Env. Sci. Tech.* **21**: 349–357.

Schulten, H.-R., and Lattimer, R.P. 1984. Application of mass spectrometry to polymers. *Mass Spectrom. Rev.* **3**: 231–315.

Wilson, M.A. 1987. NMR Techniques and Applications in Geochemistry and Soil Chemistry. Oxford: Pergamon.

Windig, W., and Meuzelaar, H.L.C. 1984. Non-supervised numerical component extraction from pyrolysis-mass spectra of complex mixtures. *Analyt. Chem.* **56**: 2297–2303.

REFERENCES FOR NEW READERS

Hayes, M.H.B., and Swift, R.S. 1978. The chemistry of soil organic colloids. In: The Chemistry of Soil Constituents, eds. D.J. Greenland and M.H.B. Hayes, pp. 179–320. Chichester: Wiley.

Hayes, M.H.B.; Malcolm, R.L.; and Swift, R.S.; eds. 1987. Humic Substances II: In Search of Structure (International Humic Substances Society). Chichester: John Wiley & Sons, in press.

Schnitzer, M., and Khan, S.U., eds. 1978. Soil Organic Matter. Amsterdam: Elsevier.

Stevenson, F.J. 1982. Humus Chemistry. New York: Wiley.

Thurman, E.M. 1985. Organic Geochemistry of Natural Waters. Dordrecht, Netherlands: Nijhoff/Junk.

APPENDIX 1 – MATHEMATICAL FORMULAE

$$\overline{P} = \frac{\Sigma P_i w_i}{\Sigma w_i} \tag{1}$$

$$\overline{C} = \frac{\text{mass of total C}}{\text{total mass}} = \frac{\Sigma C_i m_i}{\Sigma m_i} \tag{2}$$

$$\overline{r} = \frac{\Sigma r_i n_i}{\Sigma n_i} \tag{3.1}$$

where n_i is the number of moles of the ith component.

$$\overline{V} = \frac{\Sigma V_i n_i}{\Sigma n_i} = \frac{4\pi}{3} \frac{\Sigma r_i^3 n_i}{\Sigma n_i} \tag{3.2}$$

Thus $\overline{V}$ differs from $(4\pi/3)\overline{r}^3$.

$$\frac{d[CHCl_3]}{dt} = k_{obs} \sum [HS]_i \tag{4.1}$$

where $[HS]_i$ is the activity of the ith reactive site.

$$\frac{d[CHCl_3]}{dt} = \sum k_i [HS]_i \tag{4.2}$$

where k_i is the pseudo-first-order rate constant of the ith reactive site.

It follows that

$$k_{obs} = \frac{\Sigma k_i [HS]_i}{\Sigma [HS]_i}$$

APPENDIX 2 – ABBREVIATIONS

CI	– chemical ionization	GC	– gas chromatography
DEAE	– diethylaminoether	HPLC	– high performance liquid
EI	– electron impact		chromatography
FAB	– fast atom bombardment	MS	– mass spectrometry
FD	– field desorption	NMR	– nuclear magnetic
FI	– field ionization		resonance

Humic Substances and Their Role in the Environment
eds. F.H. Frimmel and R.F. Christman, pp. 165–178
John Wiley & Sons Limited
© S. Bernhard, Dahlem Konferenzen, 1988.

Binding and Transport of Metals by Humic Materials

James H. Weber

Dept. of Chemistry, Parsons Hall
University of New Hampshire
Durham, NH 03824, U.S.A.

Abstract. This review critically analyzes (a) separation and nonseparation methods used to speciate metal ions in the presence of humic matter and (b) calculations of conditional stability constants and complexing capacities for humic matter–metal complexes from the resulting data. A discussion of the nature of metal ion binding sites in humic materials follows. The last major topic is the important role of humic matter in geological transport of metals in aquatic systems. Finally, I make recommendations for future research.

INTRODUCTION TO SPECIATION TECHNIQUES

In the context of this paper *speciation*, the analytical technique of distinguishing different forms of a compound, means quantitatively differentiating free or hydrated metal ions from those bonded to humic material. Researchers have two main reasons for studying interactions of humic material with metal ions: to understand its chemical nature and to understand its effects in the environment. In my opinion we can never clarify the complex mixture of compounds in humic material by metal ion speciation studies. However, such studies are very important for prediction of the reactivity, toxicity, and transport of metal ions in the aquatic environment.

There are several criteria for development of a quality method of speciation. First, the physical principles behind the method must be sound. Second, side reactions such as adsorption, complexation, or precipitation must be controlled or minimal. Third, measurements must occur without perturbing the system. Finally, researchers should be able to measure a variety of metals of interest at appropriate low concentration levels. Compromises are necessary, because no approach completely fulfills the criteria.

There are two general philosophies toward developing and using speciation techniques. Technique-oriented researchers emphasize application of a favorite instrument or technique. Problem-oriented researchers seek solutions to problems using any approach or using two or more methods. Because of the complexity of humic material and our sparse knowledge of its chemical components, the second approach is by far the superior one.

SPECIATION BY SEPARATION TECHNIQUES

Speciation by separation approaches has two major disadvantages: there is always the possibility of shifting equilibria and of adsorbing metal ions, humic material, and their complexes on a membrane or chromatographic material. The greatest advantage of separation experiments over non-separation ones is that separation experiments allow measurement of nearly any metal ion by sensitive methods such as atomic absorption, fluorescence, and plasma spectrometry; or by nuclear chemistry techniques. This rich variety of methods means that nearly any metal ion is measurable at nano-molar concentrations. De Mora and Harrison (1983) critically reviewed separation techniques for trace metal speciation studies.

Chromatographic techniques

Chromatographic experiments use either materials which bind metal complexes or those which bind metal ions. Metal complex binding materials include hydrophobic polystyrene (XAD) resins and octadecylsilane bonded to silica (C-18), which is commonly used in reversed phase liquid chromatography. Metal-binding materials include size-exclusion materials such as Sephadex, strong cation exchange and chelating resins, and manganese dioxide. Donat et al. (1986) compared results for speciation of Cu^{2+} and other metals in ocean waters by voltammetry and extraction of humic complexes using C-18 cartridges. They found that the chromatographic technique underestimated complexed copper in surface seawaters relative to the voltammetric approach. Further investigations proved that, contrary to past assumptions, some metal complexes in surface seawater passed through the column. However, the often effective C-18 material can be cautiously used for many water samples if it binds all metal complexes.

Metal-binding chromatographic approaches also have serious drawbacks. A major problem with size exclusion chromatography is probable adsorption of humic material as evidenced by retardation of its elution by salts or buffers in the elution medium (Hine and Bursill 1984). In addition, any buffer used for elution necessitates precise knowledge of its stability constant(s) with the metal ion for correction of metal ion not bound to humic matter. Strong cation exchange and chelating resins have similar problems. Major problems with the MnO_2 metal-binding method (Van den Berg 1984) are probable adsorption of humic material and its complexes and the need to correct concentrations of free metal ion for adsorption on MnO_2 and binding in inorganic complexes.

Membrane techniques

Experiments using ultrafiltration (UF) and dialysis membranes have several problems: (a) membranes are generally expensive, (b) pores can partially clog during an experiment changing their size, and (c) metal ions, humic material, and their complexes can adsorb on membranes resulting in side reactions. Major advantages are their applicability to unmodified natural water samples and speciation of nearly any metal ion at nanomolar concentrations.

Staub et al. (1984) have thoroughly evaluated the UF method for metal ion speciation. A major prerequisite for UF studies, that metal ions pass through the membrane and that humic material and its complexes do not, is fulfilled for a fulvic acid sample. Furthermore, metal ion adsorption problems were minimal and equilibria shifts were unimportant when filtering only a small fraction of the sample. Major disadvantages for the method were that the total metal:free metal ratio should be in the 1 to 50 range and that electrolyte concentration requirements precluded use of one brand of membrane with freshwater *and* ocean samples.

Weber and co-workers (Saar and Weber 1982) studied speciation of Cu^{2+} and Cd^{2+} in the presence of soil-isolated fulvic acid and freshwater samples by equilibrium dialysis. They demonstrated the validity of the approach by several preliminary experiments. Less than 10 per cent of this fulvic acid's complexing capacity passed through the membrane, low micromolar concentrations of both metal ions equilibrated within 24 h, and a study with EDTA resulted in the correct complexing capacity. Two major problems with the experiments were that slopes for free metal ion vs. total metal ion curves well past the titration endpoint were less than unity for Cu^{2+}, and malachite $(Cu_2(OH)_2 CO_3)$ formed during addition of Cu^{2+} to one river sample. The reasons for the nonunity slopes are unknown, but the answer might be the formation of an inorganic species like malachite or adsorption of Cu^{2+} on fulvic acid–copper colloids. Dialysis experiments should be done concomitantly with light scattering studies (see below) to demonstrate the presence or absence of colloidal material.

The use of anionic membranes might lead to a renaissance of the dialysis approach. Cox and co-workers (1984) learned that metal ions equilibrated across cation exchange membranes in less than 1 h. An additional advantage of the technique was the use of inexpensive, throwaway cells. This promising technique is worthy of further study.

NONSEPARATION EXPERIMENTS

Separation and nonseparation methods share some common problems. Adsorption probably occurs with all methods in which a membrane, chromatographic material, or solid state or mercury electrode contacts a solution containing humic material. Therefore, adsorption is a potential problem with ion selective electrode and voltammetric nonseparation techniques. In addition, shifts in equilibria typical of chromatographic experiments also occur for voltammetric and ion selective electrode nonseparation ones. Nonseparation methods, unlike separation ones, are appropriate only for limited metal ions. Typical examples are Cu^{2+}, Cd^{2+}, Pb^{2+}, and Zn^{2+} with voltammetry; Cu^{2+}, Pb^{2+}, Cd^{2+}, and Hg^{2+} with ion selective electrodes; and Cu^{2+} with fluorescence spectrometry.

Ion selective electrode potentiometry

Ion selective electrodes (ISE) cannot generally be used to determine free metal ion concentrations in natural waters for several reasons. First, ISE detection limits of ca. 1 μmol/l are too high. Second, in seawater a Cu° surface impurity on the Cu^{2+} ISE reduces Cu^{2+} to Cu^{+}. Third, at concentrations below ca. 100 μmol/l adsorption–desorption processes of Cu^{2+} at the electrode surface are important. Fourth, ISEs respond to increasing concentrations of many ligands, including humic material. The latter effect for Cu^{2+} ISEs is thought to be due to dissolution of CuS from the electrode surface. Fifth, adsorption of SFA on the Cu^{2+} ISE caused slopes of free Cu^{2+} vs. total Cu^{2+} titration curves past the endpoint to decrease with increasing SFA concentration. This result suggests that calibration curves should be done in the presence of FA with excess metal ion. For all of these reasons ISEs should be used with caution.

Interesting observations result from combined ISE–light scattering experiments. Aggregation of 32 mg/l soil-isolated fulvic acid at pH 4 and 5 that occurred in the presence of 50 μmol/l Pb^{2+} (Saar and Weber 1982) is caused by neutralization of carboxylate groups by Pb^{2+}. The aggregates, which appeared well before the 400 μmol of carboxylic and phenolic ligand groups were saturated with Pb^{2+}, bound considerable Pb^{2+} by noncomplexing processes. Similar studies with fulvic acid and Cu^{2+} demonstrated that aggregation increased very rapidly after occupation of fulvic acid binding sites (Saar and Weber 1982; Underdown et al. 1985). In general, aggregation of fulvic acid decreases in the order $Pb^{2+} > Cu^{2+} \gg Cd^{2+}$.

Differential pulse anodic stripping voltammetry (DPASV)

DPASV has one major advantage for speciation of measurable metal ions in the presence of humic material: the method's sensitivity is ca. 10^{-11} mol/l, which is sufficiently low for most applications in natural water samples (Florence 1986). This advantage is overwhelmed by serious disadvantages. First, electrolytes must be added to many samples. Second, adsorption of surface-active humic material on the mercury electrode alters deposition and stripping currents in unpredictable ways, and measurement of free metal ion concentrations, except in seawater samples having low surfactant concentration, is unreliable. Third, the deposition step measures not only free metal ion but metal dissociated from complexes during the process. This process leads to erroneously high "free" metal ion concentrations. Corrections for the resulting kinetic current are difficult and unreliable. Despite potential means of overcoming these disadvantages (Florence 1986), I recommend voltammetric techniques only for seawater samples.

Fluorescence spectrometry (FS)

Isolated humic material and humic material in natural waters fluoresce. The excitation maximum is 350–360 nm and emission occurs between 420 and 480 nm. Paramagnetic metal ions, those with unpaired electrons, such as VO^{2+}, Cr^{3+}, Mn^{2+}, Fe^{3+}, Co^{2+}, Ni^{2+}, and Cu^{2+} effectively quench the fluorescence. Cu^{2+} is usually studied because it quickly forms complexes of high stability with humic material. Typically, metal ions are titrated into solutions containing humic material (Saar and Weber 1982; Ryan et al. 1983; Cabaniss and Shuman 1986).

The first assumption of the FS method is that fluorescing molecules are representative of all molecules in the humic material mixture; no one has justified this assumption, and it is likely that only a small fraction of the molecules fluoresce. Second, uncomplexed metal ions do not substantially quench fluorescence. This assumption is correct for our soil-isolated fulvic acid sample because uncomplexed metal ions at pH 1.4 do not quench its fluorescence (Saar and Weber 1982). Third, quenching by metal ions is proportional to their complexation. Calculations from FS titration data (Saar and Weber 1982) gave tyrosine–Cu^{2+} conditional stability constants and total tyrosine concentration values that agreed well with known values. This result validated the assumption for a model ligand. Furthermore, simultaneous FS and ISE studies with Cu^{2+} (Saar and Weber 1982; Cabaniss and Shuman 1986) demonstrated that Cu^{2+} binding and fluorescence quenching agree over more than four magnitudes of added Cu^{2+}.

FS studies have several advantages over most other speciation methods. First, FS studies, unlike DPASV and ISE ones, are appropriate for unmodified natural water samples. The technique is very sensitive in all media and no added electrolyte or buffer is necessary. Second, unlike techniques involving separations or electrodes, no adsorbing material must be added to samples. Third, FS conveniently allows simultaneous monitoring of light scattering (Ryan et al. 1983), which measures formation of colloidal materials containing metal ions and humic material. This auxiliary experiment is very important because the colloids adsorb metal ions, and bound metal ions will include solution and nonsolution processes (Saar and Weber 1982). Fourth, FS is the only method that measures ligand concentrations. For this reason FS is very effectively used in conjunction with a technique such as ISE that measures free metal ion concentration.

Cabaniss and Shuman (1986) used a combined FS–ISE approach for titration of Cu^{2+} into a fulvic acid sample isolated from a lake. They assumed that ligand bound to Cu^{2+} is proportional to fluorescence quenching (assumption 3 above) and determined the proportionality constant A from combined FS–ISE measurements. That is, they calibrated the relative FS measurement to the absolute ISE one and converted quenching to bound ligand without curve fitting. A is constant over a limited range of data. Three advantages of the combined FS–ISE experiments are data sets with a wider free Cu^{2+} range and lower error, avoidance of errors due to inorganic-metal complexes, and applicability to seawater matrices where the Cu^{2+} ISE cannot be used.

Recommendations for speciation studies

Because of the complex and mysterious nature of humic material I recommend experiments involving two or more simultaneous measuring techniques. Most appropriate are measurement of binding from the point of view of ligand *and* metal ion. Since FS is the most appropriate method for ligand measurement, it should be one of the chosen approaches. Probably measurement of metal ion should be done by ISE (at higher concentrations), ultrafiltration, or dialysis. These three methods, as discussed above, have the fewest and most easily overcome problems. A wide metal ion concentration range is very helpful when treating the data. In addition, because attempts are usually made to measure only solution equilibria, scattering measurements should be done to demonstrate the absence of colloidal material. When the scattered light reaches a pre-chosen level, additional data should not be used or used with the knowledge that adsorption processes probably contribute to metal ion binding.

CALCULATION OF COMPLEXING CAPACITIES AND CONDITIONAL STABILITY CONSTANTS

Sposito (1986) reviewed in detail a variety of models used to calculate binding of metal ions to humic materials and noted that the quasiparticle model offers advantages of conceptual simplicity and calculational convenience. A quasiparticle model is a mathematical description of an aqueous system containing humic material in which the actual mixture of compounds is replaced by a set of hypothetical, average, noninteracting molecules whose behavior mimics that of the actual mixture. No rigorous thermodynamic or molecular significance can be attributed to parameters calculated with the model. Quasiparticle models include affinity spectrum, polyelectrolyte, Scatchard, Perdue–Lytle, Gamble, and Buffle models. The Scatchard model with two or three binding sites very effectively models experimental data. For example, Giesy et al. (1986) tested a variety of models for calculating complexing capacities and conditional stability constants from data obtained by titration of surface waters with Cu^{2+}. They concluded that the two-site Scatchard calculation effectively modeled their data.

However, there is considerable doubt that extrapolations using Scatchard plots yield the total number of binding sites unless they are more than 50% saturated during the experiment (Klotz 1982). Binding experiments with humic material rarely or perhaps never reach the 50% level. For example, experiments with 10 mg/l soil fulvic acid (Saar and Weber 1982) having ca. 80 μmol/l carboxyl content usually result in complexing capacities of less than 30% saturation. Thus measured complexing capacity values should be used with caution.

Three warnings are relevant for use of quasiparticle models. First, parameters originating from them reflect average properties and are more useful for metal speciation calculations than for understanding metal–humic matter binding. Second, these models are not useful for predictions or for any experimental conditions except those under which the data were collected. Third, publication of mathematically manipulated data, rather than raw data, makes it impossible for others to use the data.

NATURE OF METAL ION BINDING

Most attempts to understand the nature of the metal-binding sites in humic materials involve infrared, Mössbauer, and electron paramagnetic resonance (EPR) spectroscopy. Infrared studies, even those with Fourier transform instruments, have not been very informative; researchers can usually distinguish free from metal-bound carboxyl groups, but obtain little additional information. The major metal of interest applicable for the Mössbauer technique is iron, but conclusions from the studies are sparse. For example, researchers can distinguish one or two binding sites, but no detailed information about them. EPR is the only one of the three methods with promise.

Electron paramagnetic resonance studies

Senesi et al. (1985) studied Cu^{2+} binding to a fulvic acid sample isolated from soil and learned that EPR spectra varied considerably with Cu/fulvic acid molar ratios. At high Cu^{2+} loading, the signal was unresolved because of excess aqueous Cu^{2+}, but at low loading the resolved signal indicated the donor atoms, e.g., four oxygen or two oxygen plus two nitrogen, bound to fulvic acid. The donor atom symmetry around copper was tetragonal. Removal of excess Cu^{2+} with a strong cation exchange resin left the copper complexes nearly unchanged, but resulted in resolved spectra.

Templeton and Chasteen (1980) did a detailed study of the bonding of vanadyl ion (VO^{2+}) to two fractions (fulvic acid I and fulvic acid II) of Weber's (Saar and Weber 1982) soil fulvic acid sample. They showed that for both fractions the donor atoms are oxygen and that the sites can be modeled by phthalic acid and salicylic acid ligands. The approximate molecular weights of the complexes of fulvic acid I and fulvic acid II are 3800 and 300, respectively. VO^{2+} promotes aggregation of fulvic acid into complexes in which vanadyl ions are more than 1.2 nm apart.

Recommendations for identification of metal ion binding sites

EPR has been underutilized for understanding the nature of metal ion binding sites in humic material. A proper knowledge of the theory involved (Templeton and Chasteen 1980) can lead to estimates of molecular weights of complexes formed and approximate metal-to-metal distances in the complexes. In addition, modeling of sites by known ligands can lead to strong suggestions of their nature in complexes involving humic material.

TRANSPORT OF METALS BY HUMIC MATTER

Introduction

Metals are enriched relative to average concentrations in the earth's crust in a variety of organic-rich environmental materials including petroleum, asphalt, coal, soils, shales, aquatic sediments, and peat (Rashid 1985). For example, humics-rich peat in laboratory studies enriches uranium and iron, respectively, by factors of 10,000 and 26,000 relative to aquatic concentrations. Field results similarly showed that uranium concentration was 9,000 times higher in peat (ca. 1 mg/g) than in surrounding bog waters. The ability of humics-rich peat to concentrate metals suggests that humic material can influence the transport and ultimately the deposition of metals in natural waters.

Many metals in natural waters have solubilities much higher than expected from calculations based on the inorganic ions of the medium. Researchers widely believe that the enhanced solubility is predominantly due to their complexation with, adsorption on, and reduction by humic material. In many rivers more than 50% of soluble heavy metals are associated with organic matter (Förstner and Wittmann 1983). These interactions of metal with humic material are intimately involved with their aquatic migration and, therefore, their high sediment concentrations when conditions favoring precipitation occur. Deposition of humic material and their associated metals is very favorable at the estuarine interface between fresh and salt water. Metal concentrations in estuarine sediments containing humic material are up to thousands of times greater than in humic-free sediments.

This section initially describes the ubiquity of humic matter metal binding in the aquatic environment and then emphasizes laboratory experiments, which model processes for solubilization of metals. These model system experiments link environmental processes with the surprisingly high solubility of metals in natural waters. Finally, I will describe a general approach which will lead to a better understanding of solubilizing processes induced by humic material.

Metal ion association with humic matter in natural water

Concentrations of total organic matter and humic matter in natural waters vary considerably with the source of the water (Thurman 1985). Typical total organic carbon values are 0.5 μg/ml in seawater, 7 μg/ml in rivers, and 25 μg/ml in marshes and bogs. Estuarine values vary between river and seawater values. Humic matter carbon is about 20% of dissolved organic carbon in seawater, 60% in rivers, and 70% in wetlands. These data show that humic matter is an important, if not the predominant, metal sorbing moiety in natural waters. The following discussion will not generally distinguish dissolved from particulate organic matter or adsorbed from complexed metal ions associated with humic material. I will use the general terms *sorption* or *binding* to describe the interactions.

The total binding capacity of humic acid for metal ions is 200 to 600 μmol/g (Rashid 1985). Approximately one-third of the total is cation exchange sites and the remainder is complexing sites. Humic acid surface area of about 2000 m^2/g is much greater than that of clays or metal oxides. Complexation of metals by humic material carboxylate groups is very important in solution processes, because their esterification reduces metal uptake by humic material. In addition lower molecular weight humic material fractions, which have a higher concentration of phenolic and carboxylic groups, are most effective in complexing metal ions.

Several soluble metals such as iron, mercury, copper, nickel, vanadium, and lead in natural waters are primarily associated with organic matter. For example, about 90% of soluble iron is bound to humic material, and its solubility is about 10^9 to 10^{10} greater than expected from solubility product constants. Approximately 50 to 90% of mercury in estuaries and coastal seawaters is bound to organic matter. The strongest binding is to the smallest

(<500) molecular weight cutoff fraction. Although most soluble heavy metals in natural waters are bonded to humic matter, their concentrations are not sufficiently high enough to saturate its binding sites. In rivers about 80% of humic matter binding sites are bound to Ca^{2+}, and in estuaries most sites are bonded to Ca^{2+} and Mg^{2+} (Mayer 1985).

Processes responsible for solubilization and transport of metals in the aquatic environment are related processes of dissolution of inorganic solids and prevention of precipitation through binding of metals by humic material, and increased metal ion solubility by their reduction. Illustrative laboratory experiments described below include the roles of humic material on the solubilization of a variety of inorganic compounds, enhanced solubility of metal ions in natural waters, and reduction of several metal ions to more soluble oxidation states.

Reduction of metals by humic matter

Humic matter in laboratory studies reduces a variety of metal ions including V(V) to V(IV), Hg(II) to Hg(0), Fe(III) to Fe(II), Cr(VI) to Cr(III), and U(VI) to U(IV) (Thurman 1985). In addition, humic material catalyzes photoreduction of Fe(III) to Fe(II); oxygen causes the reverse reaction. Esterification of carboxyl groups reduced the oxygen consumption by 50% thus demonstrating the importance of complexation in the process.

Solubilization of inorganic compounds by humic matter

Baker (1973) in classic experiments circulated 1 mg/ml humic acid solutions past a large variety of minerals and metals for 24 h and compared solubilized metal ion concentrations to those in a H_2O/CO_2 control solution. With rare exceptions the control solution extracted only a few micrograms of metals. In contrast, humic acid extracted more than 1 mg of metal from a variety of solids including galena (PbS), pyrolusite (MnO_2), calcite ($CaCO_3$), malachite ($Cu_2(OH)_2CO_3$), iron metal, lead metal, copper metal, and zinc metal. In some cases humic acid was superior to the synthetic ligands salicylic acid, oxalic acid, pyrogallol, and alanine; but generally its solubilizing power was similar. Humic acid also extracted metals from silicates, but usually less effectively. In addition, a six week extraction with 5 mg/ml humic acid solubilized 400 µg of silver metal and 20 µg of gold metal from coarse-grained samples.

Laboratory studies have also demonstrated solubilization of a variety of metal sulfides and carbonates by humic acid. Baker (1973) extracted Pb(II), Zn(II), Cu(II), Ni(II), Fe(III), and Mn(IV) sulfide precipitates for 30 min with a 1 mg/ml humic acid solution. In all cases the humic acid solution dissolved much more precipitate than the H_2O/CO_2 control. Solubilization by humic acid ranged from 2100 μg for Pbs to 95 μg for ZnS. In similar experiments Rashid (1985) found that humic acid at pH 7 partially solubilized all sulfides and carbonates of Co(II), Cu(II), Ni(II), and Zn(II) except for NiS. The difference in the two results for NiS probably originates from its particle size and aging.

Rashid (1985) also considered the effect of a 124 mg/ml humic acid solution at pH 8.5 on precipitation of Mn(II), Ni(II), Cu(II), and iron sulfides and carbonates. The micromoles of metal ions required for precipitate formation were 3 to 43 times more in the presence of humic acid than in its absence. Iron showed the largest effect with the carbonate and sulfide having enhanced solubility factors of 26 and 43, respectively. This result helps explain why such a high fraction of soluble iron is associated with organic matter in natural waters.

Competition studies involving metals and humic matter

Rashid (1985) studied competition among Mn^{2+}, Co^{2+}, Ni^{2+}, Cu^{2+}, and Zn^{2+} for sorption on humic acid. The range of metals bound was 98 to 150 mg/g of humic acid, and the extent of bonding decreased in the order: $Cu^{2+} \gg Zn^{2+} > Ni^{2+} > Co^{2+} > Mn^{2+}$. Then Rashid freeze-dried the metal ion–humic acid complexes and extracted a fraction of each separately with iron(III) chloride, ammonium acetate, or EDTA. All three reagents desorbed metal ions approximately as expected by the above binding order. For example, each reagent desorbed Cu^{2+} least effectively of the metal ions. The desorbing ability of the reagents was: EDTA$\geqslant$iron(III) chloride~ammonium acetate. The 78 to 98% desorption of metal ions by 0.1 mol/l EDTA demonstrated that it can readily displace them from humic acid under certain experimental conditions. In similar experiments Rashid proved that peat (ca. 20% humic material) effectively sorbed the same five metal ions from aqueous solutions; Cu^{2+} was preferentially sorbed using about 50% of the binding sites. Peat also sorbed zinc, copper, and iron from seawater in the 2 to 29 μg/g peat range.

Recommendations

The above laboratory experiments generally illuminated interesting results, but shed little light on fundamental processes involved. Two major reasons are poor design for an understanding of the processes involved and resulting emphases on observations over ideas. That is, they did not answer fundamental questions such as the following. Why does humic acid dissolve inorganic solid compounds? Why does it dissolve metals? Why does it prevent precipitation of metal sulfides and carbonates? What fundmental processes are involved? How can model experiments be designed to clarify processes involving metals in natural waters?

The dissolution of metals illustrates problems with the conclusions. There is a large difference between the dissolution of Pb and Cu by a relatively weak acid. According to standard reduction potentials the proton should oxidize Pb, but not Cu, Au, or Ag. If protons from humic material do not oxidize these metals, what does? Humic material has a reduction potential range of ca. 0.5 to 0.7 volts (vs. the standard hydrogen electrode) (Thurman 1985) and is unlikely to effect the oxidations. The answer is undoubtedly a synergic process in which extremely low concentrations of metal ions are strongly complexed by humic acid, thus shifting the equilibrium toward the complexed metal ion. Researchers could distinguish effects of oxidation by protons and of complexation by comparing dissolution of metals at pH 5 by a noncomplexing buffer solution, model ligands, and humic acids.

An approximation of the effect of side reactions on the reaction of interest is fairly simple (Ringbom and Wänninen 1979). In this technique one calculates conditional stability constants, whose values are appropriate for specific experimental conditions. Consider for an example the effect of excess ligand such as fulvic acid on the solubility of CuS at pH 7. The thermodynamic solubility product constant K has the form:

$K = [Cu^{2+}][S^{2-}] = 8 \times 10^{-36}$.

At pH 7, in the presence of fulvic acid, soluble copper will be in the forms of Cu^{2+}, $Cu(OH)^+$, and FA complexes. Soluble sulfide will be predominantly H_2S and HS^-. The binding of Cu^{2+} to hydroxide ion and fulvic acid, and of sulfide ion to protons, will all increase the concentration of soluble copper. It is not difficult to calculate the side reaction coefficient for each of the side reactions from stability constants, acidity constants, pH, and fulvic acid concentration. With these side reaction coefficients one can calculate the conditional solubility product constant for the solubility of CuS. Of course the calculations, which require approximations of fulvic acid concentration and its conditional stability constant with Cu^{2+}, are somewhat crude. A rough calculation for 10^{-4} mol/l fulvic acid solution (ca. 10 mg/l) at pH 7 shows that the conditional solubility product constant would be 1×10^{-26}, i.e., increased solubility by a factor of ca. 10^9. However, even approximate results will begin to clarify in laboratory studies the reasons for solubilization of metals by humic material.

Acknowledgements. I thank the National Science Foundation Chemical Oceanography Program for partial support of this work. I am grateful to the students and postdoctoral associates mentioned in the references for providing the creative ideas, critical insights, and experimental skills that made our work possible. I thank J.J. Alberts, M.S. Shuman, and E.M. Perdue for critical review of the manuscript and helpful suggestions.

REFERENCES

Baker, W.E. 1973. The role of humic acids from Tasmania podzolic soils in mineral degradation and metal mobilization. *Geochim. Cosmochim. Acta* **37**: 269-281.

Cabaniss, S.E., and Shuman, M.S. 1986. Combined ion selective electrode and fluorescence quenching detection for copper-dissolved organic matter. *Analyt. Chem.* **58**: 398–401.

Cox, J.A.; Slonawska, K.; Gatchell, D.K.; and Hiebert, A.G. 1984. Metal speciation by Donnan dialysis, *Analyt. Chem.* **56**: 650–653.

de Mora, S.J., and Harrison, R.M. 1983. The use of physical separation techniques in trace metal speciation studies. *Water Res.* **17**: 723-733.

Donat, J.R.; Statham, P.J.; and Bruland, K.W. 1986. An evaluation of a C-18 solid phase extraction technique for isolating metal–organic complexes from Central North Pacific waters. *Marine Chem.* **18**: 85–99.

Florence, T.M. 1986. Electrochemical approaches to trace element speciation in waters. A review. *Analyst* **111**: 489–505.

Förstner, U., and Wittmann, G.T.W. 1983. Metal Pollution in the Aquatic Environment, 2nd ed., chap. C. Berlin: Springer-Verlag.

Giesy, J.P.; Alberts, J.J.; and Evans, D.W. 1986. Conditional stability constants and binding capacities for copper (II) by dissolved organic carbon isolated from surface waters of the Southeastern United States. *Env. Toxicol. Chem.* **5**: 139–154.

Hine, P.T., and Bursill, D.B. 1984. Gel permeation chromatography of humic acid. Problems associated with Sephadex gel. *Water Res.* **18**: 1461–1465.

Klotz, I.M. 1982. Number of receptor sites from Scatchard graphs: facts and fantasies. *Science* **217**: 1247–1249.

Mayer, L.M. 1985. Geochemistry of humic substances in estuarine environments. In: Humic Substances in Soil, Sediment, and Water. Geochemistry, Isolation, and Characterization, ed. G.R. Aiken, D.M. McKnight, R.L. Wershaw, and P. MacCarthy, pp. 211–232. New York: John Wiley & Sons.

Rashid, M.A. 1985. Geochemistry of Marine Humic Substances, chap. 4 and 7. New York: Springer-Verlag.

Ringbom, A., and Wänninen, E. 1979. Complexation reactions. In: Treatise on Analytical Chemistry, 2nd ed., part 1, vol. 2, eds. I.M. Kolthoff and J.E. Elving, pp. 441–477. New York: John Wiley & Sons.

Ryan, D.K.; Thompson, C.P.; and Weber, J.H. 1983. Comparison of Mn^{2+}, Co^{2+}, and Cu^{2+} binding to fulvic acid as measured by fluorescence quenching. *Can. J. Chem.* **61**: 1505–1509.

Saar, R.A., and Weber, J.H. 1982. Fulvic acid: modifier of metal-ion chemistry. *Env. Sci. Tech.* **16**: 510A–517A.

Senesi, N.; Bocian, D.F.; and Sposito, G. 1985. Electron spin resonance investigation of copper(II) complexation by soil fulvic acid. *Soil Sci. Soc. Am. J.* **49**: 114–119.

Sposito, G. 1986. Sorption of trace metals by humic materials in soils and natural waters. *C.R.C. Crit. Rev. Env. Control* **16**: 193-229.

Staub, C.; Buffle, J.; and Haerdi, W. 1984. Measurements of complexation properties of metal ions in natural conditions by ultrafiltration: influence of various factors on the retention of metals and ligands by neutral and negatively charged membranes. *Analyt. Chem.* **56**: 2843–2849.

Templeton, G.D. III, and Chasteen, N.D. 1980. Vanadium–fulvic acid chemistry: conformational and binding studies by electron spin probe techiques. *Geochim. Cosmochim. Acta* **44**: 741–752.

Thurman, E.M. 1985. Organic Geochemistry of Natural Waters, chap. 1, 10, and 11. Boston: Martinus Nijhoff/Dr. W. Junk Publishers.

Underdown, A.W.; Langford, C.H.; and Gamble, D.S. 1985. Light scattering studies of the relationships between cation binding and aggregation of fulvic acid. *Env. Sci. Tech.* **19**: 132–136.

Van den Berg, C.M.G. 1984. Organic and inorganic speciation of copper in the Irish Sea. *Marine Chem.* **14:** 201–212.

Humic Substances and Their Role in the Environment
eds. F.H. Frimmel and R.F. Christman, pp. 179–192
John Wiley & Sons Limited
© S. Bernhard, Dahlem Konferenzen, 1988.

Interaction of Humic Substances with Biota

U. Müller-Wegener

Institut für Bodenwissenschaften der Universität Göttingen
Abteilung Chemie and Biochemie im System Boden
3400 Göttingen, F.R. Germany

Abstract. The importance of humic substances for the carbon and nitrogen cycles is summarized and the effects on plants and microorganisms are described. The possibilities of interaction of the system of humic substances and fractions of this system with biota have been divided into two groups, different in principle:

1) *direct interactions* in which the humic substances act as active agents and modify biochemical reactions directly, and

2) *indirect interactions* in which the humic substances interact by their physico-chemical properties with compounds which regulate the growth of plants or microorganisms.

INTRODUCTION

Interactions of humic substances with biota in a broad sense can only be studied within the framework of the complicated cycle, which consists not only of the production of humic substances from phytogenic, animal, and microbial initial products (Fig. 1) but also of the interaction of the produced material with the living organisms. Although a multitude of interactions of humic substances or "humus" with organisms have been demonstrated statistically, these results, originating primarily from agricultural plant production, were not investigated systematically. Only in a very few cases have the effects of the humic substances been explained using reaction mechanisms.

The discussion of the extensive results of these different interactions has to be confined to a few topics to make this survey meaningful. It seems permissible to limit the discussion to the soil system, because the behavior of humic substances in aquatic environments can be based on comparable

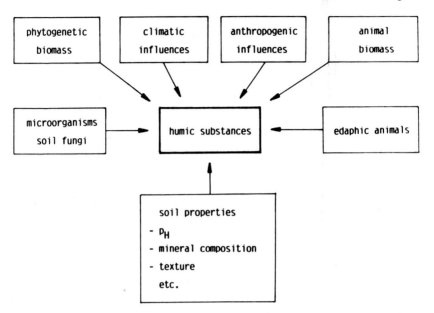

Fig. 1—Factors influencing the system of humic substances.

principles. As a matter of fact, the particular phenomena of interaction have to be weighted differently. Furthermore, the term *biota* has to be restricted to the flora and microorganisms. The problems raised by the edaphon will be left out of the discussion just as the impact of humic substances in human and veterinary medicine.

The possible interactions will be classified into two distinct categories. First, there are such interactions which directly influence the organisms involved, i.e., humic substances act as active agents and affect specific biochemical reactions. Second, indirect actions will have to be examined. In this case humic substances function, for instance, by their effect on the soil matrix or by their impact on the basic living conditions. This section also includes the interaction between humic substances and environmentally relevant chemicals.

THE SYSTEM OF HUMIC SUBSTANCES

Although humic substances do not represent a chemically uniform class of substances, these compounds show nevertheless a number of typical physicochemical properties (Ziechmann 1980). Since in most cases the influence of humic substances on phenomena involving life results from an

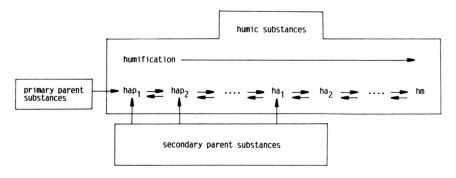

Fig. 2—System of humic substances (Ziechmann 1980; Müller-Wegener 1983); hap = humic acid precursors, ha = humic acids, hm = humine.

effect on biochemical reactions, they have to be regarded in their more differentiated structure. On the other hand, an extensive definition of humic substances should be made, which considers the lack of homogeneous structure as an important property of these compounds.

The system of humic substances may be divided into the following categories, according to experimental results (Fig. 2):

Parent substances: well-defined chemical compounds, which are transformed by humification into humic substances. Two types must be distinguished:

1) *primary parent substances*, which in most cases are compounds of aromatic character and react, for instance, by radical intermediate stages to form the skeleton of the humic substances (lignines, phenols, etc.); and
2) *secondary parent substances*, which are linked to already formed humic substances and then incorporated into their structure, losing their individuality.

Humic substances less well defined chemical compounds which can be subdivided according to their reactivity into:

1) *humic acid precursors*, substances reacting to humic acids by humification with a high reactivity,
2) *humic acids*, acids with a relatively high stability and distinct reactivity, and
3) *humines*, the final products of the humification with low reactivity and high stability.

Fulvic acids are often operationally defined as the special fraction of humic substances which is soluble in acids *and* bases. Due to their reactivity they are classified in this system as humic acid precursors.

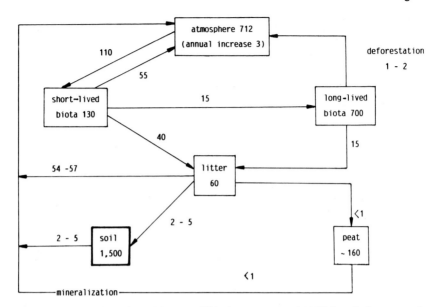

Fig. 3—Carbon cycle: size of fluxes 10^{12} kg/year, reservoir 10^{12} kg. Only conversions which are comparatively rapid (>1,000 years) are included (Stevenson 1986).

DIRECT INFLUENCES OF HUMIC SUBSTANCES ON ORGANISMS

Participation of humic substances in geochemical cycles

The relevance of humic substances to biota is particularly important in the case of two geochemical cycles, the carbon and the nitrogen cycle. For further nutrients, necessary for phytogenic and microbial life (e.g., phosphorus), the participation of the humic substance system is primarily of minor interest. Nevertheless there are some examples for reactions with humic substances in individual cases.

Carbon cycle.   Humic acids exert a direct influence on soil microorganisms by means of the carbon cycle (Fig. 3). Soil organic matter has a carbon content of about 50%. Compared to other types of organic compounds it is relatively resistant to microbial degradation. Whereas the amount of carbon which is potentially available for mineralization by microbes (1.5×10^{12} t) appears to be high, the rate of degradation amounts to only 2–5×10^9 t/year. Because the larger part of soil organic matter exists in a slightly degradable form, it is not available as a carbon source, neither can its degradation products be utilized by the microbial population or by phytogenic organisms.

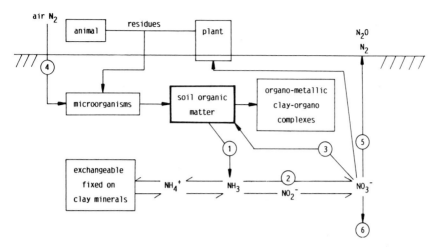

Fig. 4—Nitrogen cycle in soil: (1) ammonification, (2) nitrification, (3) immobilization, (4) fixation, (5) denitrification, (6) leaching.

The main carbon source for soil microbes is litter, of which more than 50 x 10^9 t/year are mineralized.

Nitrogen cycle. The basis of the nitrogen cycle (Fig. 4) is the mineralization of nitrogen bound in organic molecules and the immobilization of inorganic nitrogen by its conversion into organic compounds (Eq. 1).

$$\text{organic-N} \underset{\text{immobilization}}{\overset{\text{mineralization}}{\rightleftarrows}} \text{NH}_4^+, \text{NO}_3^- \qquad (1)$$

Nitrogen as a macronutrient has a decisive influence on the growth of microorganisms, plants, and, thus, on animals. It is present in most soil humic substances in different amounts (Table 1).

The nitrogen which is released by the degradation of humic substances will induce a more intensive growth in microbes and stimulate plant life via decayed microbial matter. At equilibrium the amount of released nitrogen is, of course, the same as that stemming from the pool of plant residues. In addition to the form in which nitrogen may appear, the quantitative aspect is most important. First, the ammonium and nitrate ions which are formed by microorganisms are of some interest. Second, organic nitrogen compounds are present, such as amines, amino acids, peptides, or nitrogen heterocycles. According to the type of nitrogen incorporation in the humic substance, a larger or smaller number of these compounds will be detected. The mode of nitrogen binding determines its release during degradation from the humic substance. Different mechanisms, such as charge-transfer complexes for the

Table 1 Nitrogen content of some humic acids (Müller-Wegener 1983).

Origin of Humic Acids	Total Nitrogen (%)
Synthetic humic acid (oxidized hydroquinone)	0
Black water humic acid	0.08
Lowmoor peat	1.71
Black peat	2.36
Calluna podzol B-horizon	4.78
Brown coal	5.29

initial fixation of amino acids and heterocycles (Müller-Wegener 1987) and a covalent bonding (Stevenson 1986) of amino nitrogen, have been shown to participate in the process of incorporation.

Effect of humic substances on plant growth

An analysis of the influence of humic substances on plant growth raises the question of whether these natural products of rather large molecular size are able to penetrate biological membranes. The permeation of humic substances into phytogenic organisms is essential for the participation in the metabolism of plants. For instance, it is very unlikely for a humic substance particle with a molecular weight of 100,000 to enter a root unhindered. Nevertheless, the transfer of the system of humic substances into the plant is possible. Small particles, such as humic acid precursors or humic acids, are able to penetrate biological membranes (Fig. 5, I and II) and build up a new equilibrium of humic substances in the cell (Fig. 5, III). This model has to be seen in analogy to the considerations of Ziechmann (1980) on the character of the system of humic substances.

These dualistic properties are responsible for the specific effects of humic substances on plants and, thus, allow a plausible explanation of some of the contradictory results. Not only are the influences on different plants variable, especially if parameters of a general nature such as "increase of productivity" (Table 2) are studied, but the known effects may also be reversed under different environmental conditions.

In Table 3 an incomplete summary of some interactions of humic substances or of isolated fractions of the humic substance system with plants is presented (Visser 1986). Only two results will be discussed below.

One frequently mentioned characteristic of the so-called fulvic acids (i.e., a part of humic acid precursors) is root initiation in higher plants and plant

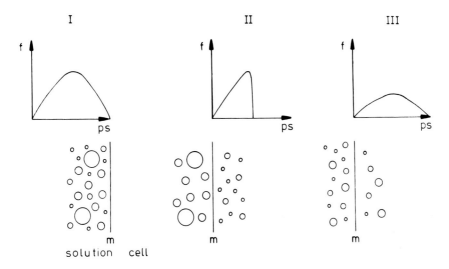

Fig. 5—A model for the transfer of a system of humic substances into living cells;
f = frequency, ps = particle size, m = membrane.

tissue cultures. It is reported that the absorption of 750 to 1,500 μg of these water soluble fulvic acids by a 5 cm hypocotyl segment of beans increases the root initiation by more than 300% (Schnitzer and Poapst 1967). Together with a negative influence of this part of the humic substances system at a concentration of 500 to 4,000 ppm on stem elongation in peas (Poapst and Schnitzer 1971), possibilities of an influence on plant growth and differentiation have been shown as well.

Excised root segments of pea seedlings showed an increase in cell elongation when treated with low concentrations of humic acids (up to 50 ppm). An increase in the concentration of humic acids over this threshold value induced a distinct decrease in elongation. While stimulating elongation, humic acids do not influence the protein metabolism, but inhibit the formation of cell wall-bound hydroxyproline from proline (Vaughan et al. 1985). If bivalent iron is added to humic acid solution, the positive effect on elongation is cancelled. Humic acids are able to bind iron ions in a chelate-like manner. The physiological effect can therefore be traced back to the decrease of the iron concentration in the nutrient solution.

These few examples obviously show that the essential effect of humic substances on higher plants is on their metabolism. To understand the macroscopic effects it is absolutely necessary to study the biochemical reactions.

Table 2 Effect of humic substances on the increase of productivity on different culti-vated plants (Khristeva and Manilova 1950).

Increase of Productivity	Plants
+++	tomato, potato, sugar beet
++	grains
+	peas, beans, lentils, cotton
nearly 0	sunflowers, castor, pumpkin

Table 3 Some effects of humic acids on plants.

Effects of Humic Acids and Related Compounds	Reference
Increase of biomass production	Kononova (1966)
Increase of stem elongation	Poapst and Schnitzer (1971)
Increase of root formation	Schnitzer and Poapst (1967)
Influence on organogenesis	Rypáček (1968)
Cytokinin-like activity	Cacco and Dell' Agnola (1984)
Inhibition of indoleacetic acid oxidase	Mato et al. (1972)
Increase of phosphatase activity	Przemeck (1962)
Inhibition of peroxidase activity	Vaughan et al. (1985)
Inhibition of phosphorylase activity	Bukvova and Tichy (1967)

Table 4 Mode of action of the humic substances on microorganisms.

Humic Acid Function	Reference
Metabolite	Gunda (1970)
Co-metabolite	de Haan (1974)
Indoleacetic acid-like molecule	Przemek (1962)
Detoxification agent	Tichy et al. (1964)
Regulator of membrane permeability	Chaminade and Blanchet (1953)
Decoupler of oxidative phosphorylation	Flaig (1970)
Respiratory catalyst	Flaig (1970)

Effect of humic substances on microorganisms

The influence on the growth of microorganisms (bacteria, fungi, and viruses) is directly connected, as already shown for plants, with humic acids penetration of membranes. A multitude of results has been reported (Table 4) demonstrating the different effects of the humic acids on microorganisms, which have a simpler structure than higher plants. In this case, too, there are partially contradictory effects for different organisms or humic acids of various origin.

Adding humic substances, at a concentration of up to 30 mg/l, to the nutrient solution generally results in an increase of the number of soil microorganisms (Visser 1985). Microbes isolated from a humus-rich soil were more stimulated than organisms from a sandy soil. On the other hand, antiphlogistic effects have been described for various humic substances (Kühnert 1979).

Effect of humic substances on enzymes

All chemical influences of humic substances on phytogenic or microbial life are ultimately actions affecting metabolism; thus they can be traced back to the interaction of humic substances with enzymes.

The inhibition or stimulation of the activity of various enzymes by humic acids has been summarized by Ziechmann (1980). Enzymes acting as exo-enzymes are of special interest for the soil system because they are found in an active form in the soil solution.

For a series of proteolytic enzymes no uniform effect by humic substances could be determined (Ladd and Butler 1975): the activities of carboxypepti-dase A, trypsin, and pronase B were inhibited, while papain, subtilopeptidase A, termolysin, and ficin were stimulated. Humic substances had no effect on tyrosin and phaseolin.

It is surprising that in these experiments no differentiation was obtained between natural humic substances and synthetic products. However, methyl-ation of humic substances caused a total inactivation (Ladd and Butler 1975). Hydrolysis of the methylated functional groups restored the activity of the humic substances.

The inhibition of trypsin and carboxypeptidase A decreased after addition of divalent cations (Ladd and Butler 1975). The mechanism will probably involve a cation exchange between the carboxyl group of the humic substance and the amino group of the protease. In no case does the formation of such enzyme-humic substance complexes imply a total inactivation of the enzyme. Active forms of complexes have been detected (Maggiano and Cacco 1977).

The results of these and further experiments (Pflug 1978) make the con-sideration of the influences of humic substances on enzymatic reactions possible, as shown in Fig. 6.

The interaction of humic substances with enzymes, therefore, can be summarized as follows:

1) Humic substances and enzymes interact directly. Binding mechanisms extend from adsorptive or steric effects to atomic bonds (e.g., an acid

amide or ester linkage). Humic substances are not only able to modify the active center by changing the quaternary or tertiary structure of the enzyme protein but can also act directly on the active sites.

2) Humic substances comprise a multitude of different structures and do not have a single molecular weight. Therefore, some humic molecules may act as analogous substrates and disturb the equilibrium of the enzymatic reaction.

3) The cation exchange properties of humic substances result in fixation of bivalent cations often used as cofactors for enzymatic catalysis or for stabilization of the structure of the protein molecule.

This last interaction leads immediately to the indirect influences of humic substances on the growth of organisms.

INDIRECT EFFECTS OF HUMIC SUBSTANCES ON BIOTA

Indirect effects of humic substances on biota influence growth or other manifestations of life not as an active agent but via interactions with other chemicals.

Binding of metal ions to humic substances

Metal cations like Fe, Cu, Zn, and Mn are micronutrients for plants and a number of microorganisms. Therefore, they greatly influence growth. Several of the transition metals are brought to plants in chelate-like form, especially

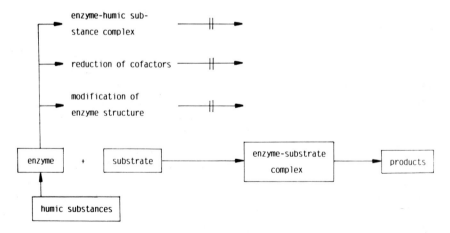

Fig. 6—Possible interactions of humic substances and enzymes.

since solubility in the soil solution at normal pH values is not high enough for a sufficient supply to several plants with, for example, uncomplexed iron. Like root exudates, humic substances are good chelating agents.

Humic substances are able to transport chelated iron towards the plant roots and besides, they stimulate the translocation of these ions into the plant. This also applies to other transition metals (Chen and Stevenson 1986). Of course, the different behavior of plants with humic substance metal complexes has to be taken into account. The different modes of uptake cause the variation in results.

Because of a better supply with micronutrients this stimulation of growth is only one partial aspect of the binding of metal ions by humic substances. The formation of soluble complexes is a precondition of the improved transport by humic substances. An effect of immobilization is caused by the significant amount of organic matter which is bound to particular soil constituents, such as oxides, clay minerals, etc. Nevertheless, there are enough functional groups which are able to bind metal ions by chelate formation. Bound metal ions are fixed by this reaction via humic substances to the soil matrix. This may cause an insufficient supply to plants.

The precipitation of compounds is another possible reaction for metal ions with humic substances which has to be considered. A number of metal ions (Cd, Pb, Hg, Ba, Ca), some of which are very relevant to the environment, form insoluble complexes (Müller-Wegener 1983) with water-soluble fractions of humic substances. The complexes thus formed are not available to plants and microbes; consequently, the concentration of toxic cations in the soil solution is reduced. This last effect leads to the most important indirect effect of humic substances on plant and microbial life.

Binding of organic chemicals relevant to environment to humic substances

More and more organic compounds come into the environment as agrochemicals or waste products from numerous syntheses. Only a few of these substances show no effect on flora and fauna. For all these compounds, which stem from anthropogenous sources, the soil system acts as a filter bed.

The filtration effect can be traced back to a number of reactions involving the main components of the soil matrix, of which humic substances play a very important part. According to the compounds present various sections of the system of humic substances will be involved in this effect, as there are, for instance, dissociated carboxylic acids, bases or nonpolar segments. Cationic substances will react with humic substances, showing a high anionic charge, by electrostatic interaction. Nonpolar compounds will react with less charged parts of the humic substances, humic acids, or even humines. Binding mechanisms range from van der Waals forces to covalent bonds (Table 5).

Table 5 Selection of some organic compounds relevant to the environment and possible mechanisms of the binding to humic substances (Hamaker and Thompson 1972; Müller-Wegener 1983).

Compound	Possibilities of Binding
s-Triazine	charge-transfer complexes
	hydrogen bonds
	ligand exchange
	covalent bonds
Amines	charge-transfer complexes
	ionic bonds
Chlorinated hydrocarbons	hydrophobic bonds
	van der Waals bonds
Polycyclic hydrocarbons	van der Waals bonds
	hydrophobic bonds
Bipyridilium cations	charge-transfer complexes
	ligand exchange
	ionic bonds
Cationic surfactants	ionic bonds

These reactions again demonstrate the different effects on biological systems. The binding of agrochemicals to humic substances, which may be of considerable magnitude, involves increased amounts of humic substances. There is no positive result on the growth of plants or microorganisms. The relative stability of the fixation to humic substances in so-called unextractable residues raises the question whether at least more or less undegradable pesticides are released by decomposition or alteration of the humic part which will subsequently damage biota.

On the other hand, fixation of environmentally relevant organic chemicals to the humic substances acts positively on several plants and microbes due to the decreased toxicity of the fixed chemicals. The decrease in concentration of these substances in soil solution depends on the amount of humic substances present.

It has to be pointed out that this property of the humic substances is of greater importance for aquatic systems. With the domestic and industrial sewage many chemicals are released into the environment, some of which are difficult to degrade. The adsorption by humic substances represents an important mechanism of detoxification.

CLOSING REMARKS

Humic substances participate very actively in the life cycles of flora and microorganisms, as demonstrated by the few examples shown above. There is no doubt that the substances of this group compete with definite organic compounds of known structure which are much more effective in their specific reaction. On the other hand, the latter compounds take part in only one type of reaction. The system of humic substances, with its immense spectrum of effects on biota, has to be looked at as a unique group of substances influencing the environment.

REFERENCES

Bukova, M., and Tichy, V. 1967. The effect of humus fractions on the phosphorylase activity of wheat (triticum aestivum L.). *Biol. Plant.* **9**: 401–406.

Cacco, G., and Dell'Agnola, G. 1984. Plant growth regulator activity of soluble humic complexes. *Can. J. Soil Sci.* **64**: 225–228.

Chaminade R., and Blanchet, R. 1953. Mécanisme de l'action stimulante de l'humus sur la nutrition minérale des végétaux. *C.R. Acad. Sci. Paris* **237**: 1768–1770.

Chen, Y., and Stevenson, F.J. 1986. Soil organic matter interactions with trace elements. In: The Role of Organic Matter in Modern Agriculture, eds. Y. Chen and Y. Avnimelech, pp. 73-116. Dordrecht: Martinus Nijhoff Publishers.

de Haan, H. 1974. Effect of benzoate on microbial decomposition of fulvic acids in Tjeukemeer (the Netherlands). *Limnol. Oceanogr.* **22**: 38–44.

Flaig, W. 1970. Effect of humic substances on plant metabolism. Proc. 2nd Int. Peat Congress, Leningrad, pp. 579–606.

Gunda, B. 1970. Der Einfluss der Humusstoffe auf die Anzahl der Bodenkleinlebewesen in Kulturen. *Zentralbl. Bakt. Parasit. Infektionskr. Hyg.* **125**: 584–593.

Hamaker, J.W., and Thompson, J.M. 1972. Adsorption. In: Organic Chemicals in the Soil Environment, eds. C.A.I. Goring and J.W. Hamaker, pp. 49–143. New York: Marcel Dekker.

Khristeva, L.A., and Manoilova, A. 1950. Zur unmittelbaren Einwirkung der Huminsäuren auf das Wachstum und die Entwicklung der Pflanzen. *Ber. Lenin-Akad. Landw. Wiss.* **1**: 10–21.

Kononova, M.M. 1966. Soil Organic Matter. Oxford: Pergamon Press.

Kühnert, M. 1979. Untersuchungen über chemische Eigenschaften sowie chemisch-toxikologische und pharmakologisch-toxikologische Wirkungen von Huminsäuren mit der Zielsetzung ihrer Anwendung in der Medizin (speziell Veterinärmedizin). Promotion B, Karl-Marx-Universität, Leipzig.

Ladd, J.N., and Butler, J.H.A. 1975. Humus–enzyme systems and synthetic organic polymer–enzyme analogs. In: Soil Biochemistry, eds. E.A. Paul and A.D. McLaren, vol. 4, pp. 143–194. New York: Marcel Dekker.

Maggiano, A., and Cacco, G. 1977. Acetylnaphtylesterase activity in humus–enzyme complexes of different molecular size. *Soil Sci.* **123**: 122.

Mato, M.C.; Olmedo, M.G.; and Méndez, J. 1972. Inhibition of indoleacetic acidoxidase by soil humic acids fractionated on Sephadex. *Soil Biol. Biochem.* **4**: 469–473.

Müller-Wegener, U. 1983. Neue Erkenntnisse zur Wechselwirkung zwischen s-Triazinen und organischen Stoffen in Böden. Habilitationsschrift. Georg-August-Universität, Göttingen.

Müller-Wegener, U. 1987. Electron donor acceptor complexes between organic nitrogen heterocycles and humic acids. *Sci. Total Env.* **62**: 297–304.

Pflug, W. 1978. Über Aktivitätsveränderungen ausgewählter hydrolytischer Enzyme durch Huminstoffe. Dissertation. Georg-August-Universität, Göttingen.

Poapst, P.A., and Schnitzer. M. 1971. Fulvic acid and adventitious root formation. *Soil Biol. Biochem.* **3**: 215–219.

Przemek, E. 1962. Physiologische und histochemische Untersuchungen über Einflüsse von synthetischen und natürlichen Huminstoffen auf das Wurzelwachstum. Dissertation. Georg-August-Universität, Göttingen.

Rypácek, V. 1968. Humic acids as related to plant morphogenesis and toxicity. *Pontif. Acad. Sci. Scri. Varia* **32**: 725-729.

Schnitzer, M., and Poapst, P.A. 1967. Effects of a soil humic compound on root initiation. *Nature* **213**: 598-599.

Stevenson, F.J. 1986. Cycles of Soil. New York: Wiley-Interscience.

Tichy, V.; Mannsbartová, E.; and Minár, J. 1964. Über die Natur des Entgiftungseffektes der Humussäuren. *Biol. Plan.* **6**: 306–314.

Vaughan, D.; Malcolm, R.E.; and Ord, B.G. 1985. Influence of humic substances on biochemical processes in plants. In: Soil Organic Matter and Biological Activity, eds. D. Vaughan and R.E. Malcolm, pp. 77–108. Dordrecht: Martinus Nijhoff Publishers.

Visser, S.A. 1985. Physiological action of humic substances on microbial cells. *Soil Biol. Biochem.* **17**: 457–462.

Visser, S.A. 1986. Effects of humic substances on plant growth. In: Humic Substances, Effects on Soil and Plants, REDA, pp. 89–135. Rome.

Ziechmann, W. 1980. Huminstoffe. Weinheim: Verlag Chemie.

Humic Substances and Their Role in the Environment
eds. F.H. Frimmel and R.F. Christman, pp. 193–214
John Wiley & Sons Limited
© S. Bernhard, Dahlem Konferenzen, 1988.

Environmental Photoprocesses Involving Natural Organic Matter

R.G. Zepp

U.S. Environmental Protection Agency
Environmental Research Laboratory
Athens, GA 30613, U.S.A.

Abstract. Current research is reviewed on the photoreactions that occur when sunlight interacts with soil and aquatic organic matter. The primary focus is on photoprocesses involving humic substances. Investigations of the direct photoreactions of humic substances are discussed, with emphasis on the fading of the UV-visible and fluorescence spectra. The balance of this background paper deals with sunlight-induced reactions between humic substances and trace organic or metallic substrates. Recent studies have shown that these indirect photoreactions are mediated by electronically excited states or reactive intermediates derived therefrom that react with substrates by electronic energy transfer, by electron transfer, and/or by various free radical processes. Concepts and methods that are used to elucidate the nature and concentrations of these transient reactants are discussed, including approaches based on steady-state irradiations and techniques that employ laser flash photolysis.

INTRODUCTION

Natural organic matter is known to be among the most important sunlight-absorbing components of soil surfaces and aquatic environments. It has been known for some time that the dissolved organic matter in water bodies is susceptible to photochemical transformation. During the early 1900s extensive investigations demonstrated that sunlight was responsible for bleaching of the organic color in the upper layers of water bodies. This early work has been reviewed by Whipple (1914) and Il'in and Orlov (1979). Very recent studies have focused on the nature of the photochemistry that occurs when natural organic matter absorbs light. Through spectroscopic and photochemical studies, scientists have begun to realize that solar irradiation of the organic matter results in a rich variety of photochemical processes.

One of the basic principles of photochemistry is that light must be absorbed in order for photochemical reactions to occur. With this principle in mind, most of the emphasis in this and succeeding sections is on the colored organic matter that strongly absorbs sunlight. According to Thurman (1985), about half of the organic matter and nearly all of the colored organic matter in freshwaters are humic substances. For clarity in this review, the term humic substances is reserved only for fulvic acids or humic acids that were isolated from soils or water bodies. In other cases, such as studies in natural water samples, the term dissolved organic matter (DOM) is used to demonstrate that the humic substances were not isolated from the water for study.

Although this overview highlights photoprocesses involving humic substances, research discussed by Mopper and Zika (1987) indicates that other more dilute components of aquatic organic matter, such as flavins, may also make important contributions to natural water photochemistry. Moreover, a number of naturally occurring organic ligands that are transparent to sunlight can react with trace metals to form photoreactive, sunlight-absorbing complexes (Waite 1986).

The reader who is interested in further delving into the area of natural water photochemistry is referred to a recent review by Zafiriou and co-workers (1984). For lead references on the effects of dissolved organic matter on penetration of sunlight into water bodies, the reader should see the recent paper by Baker and Smith (1982).

This paper first covers the effects of sunlight on the alteration of humic substances. Then, various indirect photoreactions that are initiated through light absorption by organic matter are discussed. (An indirect photoreaction is a light-induced reaction of a substrate that is initiated through light absorption by another chemical in the system (Zafiriou et al. 1984). Such reactions are sometimes generally referred to as "photosensitized" although some photochemists have reserved this term for indirect photoreactions involving energy transfer.) The discussion of indirect photoreactions covers reactions involving electronic energy transfer, electron transfer, and free radical formation. The concluding section discusses photoreactions of complexes of natural organic matter with metals and with organic chemicals.

DIRECT PHOTOREACTIONS OF HUMIC SUBSTANCES

Although it is well established that the humic substances in soils are mainly derived from biological and chemical decomposition of higher plant matter, the origin of humic substances in aquatic environments is more obscure. Some evidence exists that photochemical reactions may play a role in the formation of aquatic humic substances. Harvey and co-workers (1983) have presented evidence that the weakly colored, extractable DOM in open seawater forms via light-initiated autooxidation of plankton-derived substances. Recent studies by Leenheer (personal communication) indicate that a portion of the DOM that he isolated from the Suwannee River, Georgia, is made up of hydroxyphenyl ketones that possibly were derived from a photochemical isomerization of corresponding aryl esters known as a "photo-Fries reaction" (Chapman 1967).

Spectral properties of aquatic organic matter have been used to assess its origin and concentration. The DOM isolated from the open ocean absorbs visible light much less strongly than the well-known "Gelbstoffe," which is very important in the underwater optics of coastal seawater. This difference supports the suggestion of Højerslev (1982) that a significant portion of coastal DOM must be terrestrial in origin. Indeed, Hutchinson (1975) has applied a similar argument based on spectral properties to assess the origin of lake water DOM. Allochthonous DOM is hypothesized to be highly colored whereas autochthonous DOM is postulated to be weakly colored. Such arguments based on spectral properties of aquatic DOM are compelling in their simplicity. As discussed in the balance of this section, however, this type of analysis should be regarded with some caution in view of the susceptibility of the colored DOM and its fluorescence to sunlight-induced fading.

As mentioned in the introduction, it has been long established that the organic matter from soils and in water bodies is photoreactive. Evidence for such photoreactivity derives from observed changes in the electronic absorption spectrum of the organic matter as well as in the fluorescence intensity and spectrum. Changes that we have observed in the absorption spectrum of a water sample from the Okefenokee Swamp (Fig. 1) are similar to those reported by others. The figure also illustrates the stability of the absorption spectrum of the colored organic matter in the absence of light. The intriguing phenomenon of retardation of fading in the absence of dioxygen has also been found by Gjessing (1976) and Frimmel and Bauer (1987). Both our observations and those of Gjessing (1976) indicate that the sunlight-induced decrease in absorbance was not accompanied by a corresponding change in dissolved organic carbon (DOC). On the other hand, research results reviewed by Choudhry (1981) indicate that irradiation of humic substances under acidic conditions with short-wavelength ultraviolet light results in extensive mineralization. Under neutral and alkaline conditions a soil fulvic acid photoreacted to form benzene carboxylic and aliphatic fatty acids.

Detailed studies of the light-induced changes in spectral properties of natural organic matter have been conducted by Il'in and Orlov (1979), Kouassi (1986), Kouassi and co-workers (1986), and Frimmel and Bauer (1987), among others (Choudhry 1981). These studies have shown that humic substances are faded by exposure to UV and visible light either in aqueous solutions or as solid films. Parameters proportional to quantum efficiencies

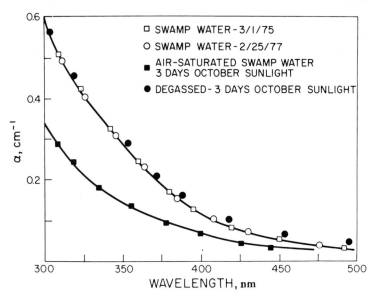

Fig. 1—Sunlight-induced fading of a water sample obtained from the Okefenokee Swamp, Georgia, U.S.A. The symbol α_λ represents the decadic absorption coefficient of the water sample at wavelength λ.

have been reported for fading of humic substances by Il'in and Orlov (254 nm) and by Kouassi and co-workers (several wavelengths characteristic of sunlight) (1986).

The most extensive studies available on photoreactions of aquatic organic matter were recently completed by Kouassi and co-workers (1986). These studies showed that in sunlight, the fluorescence intensity of humic substances fades more rapidly than the absorbance. The light-induced changes in fluorescence are wavelength dependent, and, as also shown by Frimmel and Bauer (1987), the net effect of short wavelength UV light (254 nm) is an increase in the fluorescence intensity. Also, Kouassi found that the photophysical and photochemical properties of certain coastal marine DOM and terrestrial humic substances were remarkably similar. Finally, using action spectra for the fading of DOM and computer-simulated underwater solar spectral irradiance, it was shown that observed changes in fluorescence intensity:absorbance ratios as a function of depth in seawater can be largely attributed to photochemical reactions of the DOM in the water. These studies, like most of the research on direct photoreactions of natural organic matter, include no information on the molecular structures of the chromophores or the photoproducts of the reactions. Also, little is known about the nature of the oxidizing species that are involved in the fading and the possible role of these light-induced reactions in the cycling of carbon in the environment.

In addition to direct photoreactions of the DOM, its oxidation by light-initiated free radical reactions also contributes to its degradation. These radical pathways are discussed later in the paper.

PRIMARY PHOTOPROCESSES OF HUMIC SUBSTANCES

The absorption of solar radiation by humic substances leads to the production of short-lived, electronically excited states. The excited states and reactive intermediates derived therefrom participate in a variety of reactions with organic and inorganic substrates that are present in natural waters and soil surfaces.

An abbreviated scheme that illustrates the very rapid primary processes of the electronically excited states that result on light absorption by organic substances is shown below (Eqs. 1–12). For more detailed discussions of these processes and the reactivities of different types of organic excited states, the reader is referred to Turro (1978).

The symbol "S" throughout this scheme generically represents the photoreactive components of humic substances and, in like manner, "RH" denotes the various substrates (including components of the humus itself) that participate in chemical reactions that are initiated through light absorption by humic substances. The asterisk (*) following the "S" symbol indicates that the molecule is in an electronically excited state. Because humic substances are mixtures, it is likely that excited states of several different kinds of molecules may be involved in each of the processes shown in Eqs. 1–12.

On light absorption, a molecule is promoted to its first excited state, $^1S^*$. Singlet excited states are very short-lived, about 1 nanosecond in the case of humic substances (Power et al. 1987). Excited singlet states relax back to the original ground state either by light emission (fluorescence) (Eq. 2) or by radiationless decay accompanied by heat loss (Eq. 1). Excited singlet states also decay in part by intersystem crossing to excited triplet states, $^3S^*$, which are much longer lived (Eq. 4). Like excited singlets, triplets decay by a variety of processes, but intersystem crossing to the ground state (Eq. 5) is relatively slow.

Both excited singlets and triplets can interact with other substances through bimolecular processes. These bimolecular processes can be "diffusional" or "static". In the case of diffusional processes, the excited S molecule diffuses together with some other molecule then interacts with it while they are nearest neighbors in solution. Static bimolecular processes are generally defined as interactions that occur when the chromophore is already a nearest neighbor to the other molecule when it is excited by light absorption. Static interactions are most likely to be important with soil surface photoprocesses or with complexes between the humic substance and the substrate. In this section, diffusional bimolecular processes will be emphasized. Photoreactions involving complexes of humic substances with other substances will be reviewed later in the paper.

$$S \longrightarrow {}^1S^*$$

$$\longrightarrow S + HEAT \qquad (1)$$

$$\longrightarrow S + LIGHT \qquad (2)$$

$$\xrightarrow{+\ RH} PRODUCTS \qquad (3)$$

$$\longrightarrow {}^3S^* + HEAT \qquad (4)$$

$${}^3S^*$$

$$\xrightarrow{+\ O_2} S + HEAT \qquad (5)$$

$$\xrightarrow{+\ O_2} S + {}^1O_2 \qquad (6)$$

$$\xrightarrow{+\ O_2} S+ \ + O_2^- \qquad (7)$$

$$\longrightarrow \cdot SO_2 \qquad (8)$$

$$\xrightarrow{+\ RH} I$$

$$\longrightarrow S + RH \qquad (9)$$

$$\longrightarrow PRODUCTS \qquad (10)$$

$$I = \begin{array}{l} S \ + {}^3RH^* \\ SH + R \\ S\text{-} + RH\text{+} \end{array}$$

$$S^* \xrightarrow{+\ H_2O}$$

$$\longrightarrow S+ \ + e^-_{aq} \qquad (11)$$

$$\longrightarrow SH\cdot + \cdot OH \qquad (12)$$

Because concentrations of other chemical substances in freshwaters are very dilute and excited singlets are very short-lived, the efficiency of bimolecular processes involving excited singlets in freshwater bodies is very low. Such singlet processes potentially could be significant at the much higher concentrations that occur on soil surfaces. Moreover, singlet states may undergo unimolecular decomposition to reactive intermediates such as solvated electrons and radical cations (Eq. 11), which can participate in secondary reactions with substrates.

Bimolecular processes (Eqs. 6–10) are a much more important fate for triplets than for singlets. Such processes include energy transfer to or reaction with dioxygen (Eqs. 6–8) and reactions with the substrate with subsequent secondary reactions to form products (Eq. 10) or to revert to starting materials (Eq. 9). Triplets are also potentially susceptible to physical quenching by environmental chemicals other than dioxygen or substrates, including other components of the DOM itself, but, as discussed below, research conducted to date indicates that the major fate of triplets in aquatic environments is quenching by dioxygen. The effects of added quenchers on indirect photoreaction rates can provide, however, useful information about the nature of the triplets and other transients. As with excited singlets, solvated electrons can also form triplet states (Eq. 11). Finally, there is evidence that humic substances react by some mechanism in water to form hydroxyl radicals. This reaction is shown formally as a hydrogen atom transfer in Eq. 12, consistent with the recent studies of Ononye and co-workers (1986) who suggest that triplet quinones can abstract hydrogen atoms from water.

METHODS FOR INVESTIGATING INDIRECT PHOTOREACTIONS

Two methods have been mainly used to elucidate the nature and concentrations of the excited states and reactive intermediates involved in indirect photoreactions. These reactions have been studied using steady-state irradiations and laser flash photolysis. The excited states and intermediates are together designated as transient reactants in the following section.

Steady-state irradiations

This method involves the continuous irradiation of a humus solution or natural water sample to which some trace substrate has been added. Under continuous irradiation the transient reactants rapidly reach a steady-state in which their rates of generation (v_g) equal their rates of decay (v_d). The generation and decay can be summarized by Eqs. 13–15:

$$S + LIGHT \longrightarrow X \tag{13}$$

$$X + RH \begin{cases} \xrightarrow{1/\tau_x} \text{unreactive decay product of X} & (14) \\ \xrightarrow{k(X + RH)} \text{reaction product(s)} & (15) \end{cases}$$

where X can be $^3S^*$, $^1O_2$, O_2^-, RO_2^-, e_{aq}^-, or $\cdot OH$ (Eqs. 1–12). The lifetime of the transient reactant, τ_x, is determined by its reactions with system components, including quenching by the solvent, reactions with the DOM, reaction with or quenching by dioxygen, reactions with itself (Eq. 14), and reaction with the substrate (Eq. 15). In Eq. 14, τ_x is the lifetime in the absence of RH and in Eq. 15 k(X + RH) is the second-order rate constant for reaction of X with the substrate. For a given transient reactant, the lifetime in a natural water may be controlled by only one of these reactions. The rate expression describing the formation of X is:

$$v_g = \int v_g(\lambda) \, d\lambda = \int I_a(\lambda) \, \phi_x(\lambda) \, d\lambda \tag{16}$$

where $v_g(\lambda)$ and $I_a(\lambda)$ are, respectively, the generation rate and the light absorption rate of the chromophore at wavelength λ and $\phi_x(\lambda)$ is the quantum yield for production of X. The integration is required if polychromatic light is used. The quantum yield is the fraction of light absorbed at a certain wavelength λ that results in formation of the transient reactant.

The decay rate of X is: $v_d = \{1/\tau_x + k(X + RH) [RH]\} [X].$ \hfill (17)

If RH is very dilute or unreactive towards X, then its lifetime is nearly the same as in the absence of RH and $v_d = [X]/\tau_x$. At steady state, $v_g = v_d$ and $[X]_{ss}$, the concentration of the transient reactant, is:

$$[X]_{ss} = v_g \tau_x. \tag{18}$$

The indirect photoreaction rate of the substrate (conversion per unit time) is described by:

$$-d[RH]/dt = k(X + RH) [X]_{ss} [RH]. \tag{19}$$

Under conditions in which the fractional change in chromophore concentration or in the lifetime of the transient reactant is insignificant compared to that of the substrate, the reaction is pseudo-first-order overall with a first-order rate constant, $k_{i,x}$. On the other hand, if the concentration of RH is so great that it reacts with essentially all of the transient reactant that is formed, then the photoreaction rate equals v_g and the reaction is described by a zero order rate expression.

Information about the nature and steady-state concentrations of the transient reactants can be derived from experiments using substrates to probe specifically for a certain transient reactant. Under the pseudo-first-order reaction conditions, the steady-state concentration of X can be calculated from the pseudo-first-order rate constant and from the known second-order rate constant for reaction of the transient reactant with the probe. Given the lifetime of the transient reactant, its generation rate can be computed from the steady-state concentration using eq. 18. The quantum yield for photoproduction of a transient, $\phi_X(\lambda)$, at wavelength λ, can be computed from the generation rate at that wavelength and the rate of light absorption of the solution, i.e. $\phi_X(\lambda) = v_X(\lambda)/Ia(\lambda)$.

Lead references to the application of steady-state irradiations in studies of photoreactions involving humic substances appear in recent journal articles by Zepp and co-workers (1985), Haag and Hoigné (1986), and Frimmel and co-workers (1987). Specific examples of the results derived from such studies are described later in this paper.

Laser flash photolysis

During the past two years several research groups have employed lasers to investigate the excited states and reactive intermediates that form on irradiation of humic substances. Solutions of humic substances are irradiated by a very brief, high intensity pulse of light from a laser. The formation and decay of the transients can be monitored on a time scale that ranges from picoseconds to microseconds. The excited singlet states have been monitored through detection of the fluorescent light emitted. Triplets and other intermediates are monitored through changes in the light absorption. The nature of the transient reactants can be inferred by examination of their absorption spectra and the effects of quenchers on their decay rates and quantum efficiencies for production.

For thorough discussions of this research area, papers by Fischer and co-workers (1987), Power and co-workers (1986, 1987), Milne and co-workers (1987), and Frimmel and co-workers (1987) are recommended.

PHOTOSENSITIZATION

Energy transfer processes involving humic substances are discussed by Haag and Hoigné (1986), Zepp and co-workers (1985), Fischer and co-workers (1987), Power and co-workers (1987), and Frimmel and co-workers (1987).

Dioxygen, also known as oxygen or molecular oxygen, is an excellent triplet quencher and its concentration in the photic zone of water bodies is considerably higher than that of other potential quenchers. Through analysis of the results of varying dioxygen concentrations on humus-sensitized reactions, Zepp and co-workers (1985) concluded that, in air-saturated water, the primary fate of the triplets of a variety of humic substances is quenching by dioxygen. Laser flash photolysis studies also demonstrated that transient absorption of humus triplets is strongly quenched in air-saturated solutions (Frimmel et al. 1987; Fischer et al. 1987).

The quenching process, in part, involves energy transfer to regenerate the sensitizer in its ground state and to excite dioxygen to its singlet state, $^1O_2$. In this state dioxygen is often referred to as "singlet molecular oxygen," or simply "singlet oxygen." In addition to energy transfer, the quenching of triplets by dioxygen may involve electron transfer from triplet to dioxygen to form superoxide ion. Finally, recent studies (see Gorman and Rodgers (1986) for lead references) indicate that the quenching of triplets further may involve chemical reaction with dioxygen to form biradicals. This section focuses only on the energy transfer process and singlet oxygen reactions. The other quenching processes are covered later in the paper.

Singlet oxygen studies in water

Singlet oxygen is a more potent oxidant than is ordinary dioxygen and a great deal is known about its chemical reactivity. Light-induced oxidations involving the intermediacy of singlet oxygen are referred to as photooxygenations. It was first detected in irradiated natural waters (Zepp et al. 1977) through application of the steady-state irradiation method. 2,5-Dimethylfuran (DMF) was selected as a substrate to probe for singlet oxygen in these studies. Although photostable at wavelengths > 300 nm in distilled water, DMF photoreacted rapidly in various sunlight-exposed, colored natural water samples to form oxidation products that are known to form when singlet oxygen reacts with DMF (Usui and Kamagawa 1974).

2,5-Dimethylfuron

Mechanistic tests provided further kinetic evidence that singlet oxygen was the transient oxidant involved in the DMF photooxidation. The photooxidation rate of DMF in the natural water samples was enhanced by addition of deuterium and inhibited by addition of the amine DABCO. The degree of enhancement or inhibition by these reagents was the same as observed for the photooxygenation of DMF in aqueous solutions of rose bengal, a known singlet oxygen photosensitizer.

Haag and co-workers (1984), Zepp and co-workers (1985), and Frimmel and co-workers (1987) have further shown that singlet oxygen is photogenerated in solutions of soil or aquatic humic substances as well as in natural water samples. Haag and co-workers (1984) and Frimmel and co-workers (1987) used a singlet oxygen trapping agent, furfuryl alcohol, that was first introduced by Braun's research group. Efforts to isolate the photosensitizing activity by separation of humic substances into various molecular weight fractions have had mixed success. Haag and Hoigné (1986) found little difference in the photosensitizing activity of various fractions of natural organic solutes from Swiss water bodies whereas Frimmel and co-workers (1987) found significantly higher activity in lower molecular weight fractions of Bavarian aquatic humic substances.

The effects of pH and wavelength on singlet oxygen reactions in aqueous solutions of humic substances have also been investigated. Results of Zepp and co-workers (1981) (Fig. 2) indicated that the effects of pH variation on the photooxidation of DMF in a colored river water sample were most pronounced at extreme pH values (over 9 and under 4) that usually are not found in natural water bodies. Similar results were found by Haag and Hoigné (1986) and by Frimmel and co-workers (1987) in their studies of pH effects on the photooxidation of furfuryl alcohol in DOM solutions. Such studies are scarce, however, and it would be premature to generalize these results, especially because there is no theoretical basis to do so.

The effects of wavelength on photosensitized oxygenations of DMF and the photosensitized isomerization of 1,3-pentadiene have been examined using monochromatic light. Zepp and co-workers (1981, 1985) have used "response functions" to define the action spectra for these photoreactions. The response function is defined as the ratio of the pseudo-first-order rate constant for the reaction to the average irradiance at a given wavelength. These action spectra are used to compute the depth dependence of the photooxygenations. Comparison of the action spectra to the absorption spectra of a river water sample and a solution of a commercial substance called "Fluka humic acid" (Fig. 3) indicate that: (a) the action spectra drop off more rapidly than the absorbance, indicating that the quantum yields increase with decreasing wavelength, especially at wavelengths < 360 nm; (b) visible light is much more effective at producing singlet oxygen in the "humic acid" solution than in the river water. It is interesting to note that comparison of the corrected fluorescence excitation and UV absorption spectra of an aqueous solution of a fulvic acid indicates that the quantum yield for fluorescence exhibits a wavelength dependence opposite to that

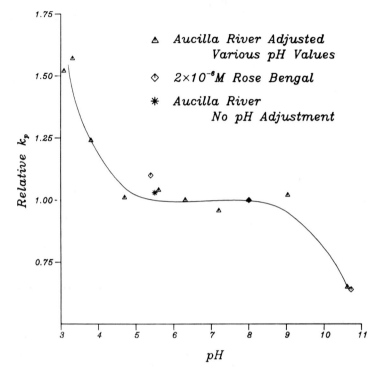

Fig. 2—Effects of pH on photooxygenation of 2,5-dimethylfuran in a colored river water sample and in aqueous solutions of rose bengal (Zepp et al. 1981).

shown in Fig. 3 (Donard et al. 1987). The fluorescence quantum yield decreases with decreasing wavelength at wavelengths < 360 nm.

For a given humic substance, near-surface, steady-state concentrations of singlet oxygen ($[_1O^2]_{ss}$) in sunlight-exposed solutions were proportional to the UV absorbance and, consequently, to the humus concentration (Haag and Hoigné 1986; Zepp et al. 1985). At 365 nm, ratios of average annual $[_1O^2]_{ss}$ to Napierian absorption coefficients (m^{-1}) of the solutions fell in the range of about $(0.5-1.5) \times 10^{-14}$ mol/l/m for the mid-northern latitudes. Lakewater DOM fell at the upper end of this range and commercial humic acid solutions and municipal wastewater at the lower end. The highly colored organic matter from swamps and soil fulvic acids photoproduced singlet oxygen somewhat less efficiently than the more weakly colored lakewater DOM, which undoubtedly is derived to a greater extent from decay of plankton.

Perhaps the most noteworthy result of these studies is the narrowness in the range observed, considering the diversity in origins of the organic matter that was studied. Malcolm and McCarthy (1986) have recently pointed out

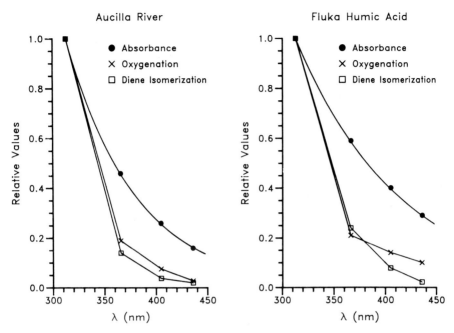

Fig. 3—Comparison of action spectra (normalized to 313-nm value) for photosensitized reactions and absorption spectra for a colored river water sample and an aqueous solution of a commercial substance called "Fluka humic acid" (Zepp et al. 1985).

that the commercial "humic acids" from Aldrich and Fluka have different properties, especially NMR spectra, compared with aquatic and soil humic substances. In this and other studies, these commercial substances were found to be somewhat less efficient in their photoproduction of singlet oxygen than were the aquatic DOM and their action spectra also differed somewhat (Fig. 3). Whether these differences are sufficient to reject these commercial materials as model photosensitizers is a matter to be judged by each investigator.

Of course, the absorption coefficients of water bodies vary over orders of magnitude and so do near-surface singlet oxygen concentrations. For coastal waters $[^1O_2]_{ss}$ about 10^{-14} have been reported, for eutrophic lakes, 10^{-13}, and for blackwater swamps $> 10^{-12}$.

The near-surface concentrations are useful for comparing photosensitizing properties of the organic matter from various sources. Another useful kinetic parameter is the quantum yield for singlet oxygen photoproduction. Under conditions that are observed in many natural water bodies, the sunlight is completely absorbed in the water column. In such a water body, the average concentration of singlet oxygen produced at a given wavelength in the water column is proportional to the quantum yield at that wavelength and inversely proportional to water depth. Studies to date indicate that about 0.4 to 3% of the ultraviolet and blue portion of sunlight absorbed by natural waters results in singlet oxygen formation (Haag et al. 1984; Zepp et al. 1977, 1985; Haag and Hoigné 1986; Frimmel et al. 1987).

Results of Haag and Hoigné can be used to estimate that, for several Swiss lakes, the average $^1O_2$ concentrations (mol/l) in water columns that have completely absorbed the solar UV light are about $5 \times 10^{-14}/z$, where z is the depth of the water column in meters. As discussed above, the pseudo-first-order rate constant for photooxygenation of a substrate equals $k_A[^1O_2]_{ss}$, where k_A is the second-order rate constant for reaction of singlet oxygen with the substrate. Extensive compilations of k_A values are available and are referred to in the papers mentioned at the beginning of this section. As examples, the computed half-life for oxygenation of DMF in the top 10 meters of a Swiss lake is about 1 week. In the highly colored Okefenokee Swamp in south Georgia (depth about 0.5 meters), the half-life would be about 4 hours.

It should be kept in mind that singlet oxygen is a highly selective oxidant, reacting rapidly only with electron-rich compounds such as furans, certain olefins, polycyclic aromatic hydrocarbons, sulfides, amines, and phenolate ions. Some amino acids and proteins are readily oxidized by singlet oxygen and it is toxic to many microbiota. The ecological significance of photosensitized formation of singlet oxygen in natural waters is poorly understood, however.

Singlet oxygen reactions on irradiated soil surfaces

Comparatively little is known about photoreactions of chemicals on soil surfaces. Gohre and co-workers (1986) recently have provided evidence that singlet oxygen may be photoproduced on soil surfaces. Two alkenes were used to trap the singlet oxygen on surfaces that were exposed to sunlight or UV lamps. Products that were characteristic of singlet oxygen reactions were found in these experiments along with products derived from other types of oxidizing transients. Analysis of the kinetic results indicates that steady-state concentrations of $^1O_2$ on sunlight-irradiated soil surfaces were comparable in magnitude to the highest singlet oxygen concentrations that have been measured in natural waters. Because no correlation was found between the photooxygenation rate constants and soil organic content, it was concluded that other components of the soil were primarily involved in the singlet oxygen photoproduction. Another possible interpretation of the results, however, is that the soil organics were mainly responsible for $^1O_2$ production, but that the organic matter also was primarily responsible for retarding light attenuation and/or quenching of the singlet oxygen (or excited state precursors) that was produced.

Photoreactions involving energy transfer to substrate

Zepp and co-workers (1985) have used the photoisomerization of 1,3-penta-diene shown below (also discussed by Turro 1978) to elucidate the nature and concentrations of triplet states that form on irradiation of humic substances in water.

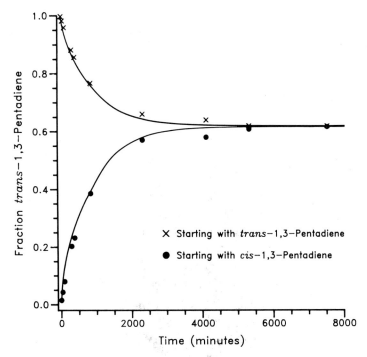

Cis-1,3-Pentadiene   Trans-1,3-Pentadiene

Kinetic results of these studies are illustrated by Fig. 4. On irradiation of either the *cis* or *trans* isomer in an air-saturated solution of a humic substance or in a colored natural water sample, the diene isomerizes until the rate of *cis* formation equals the rate of *trans* formation. At this point the reaction is said to have reached a "photostationary state" in which the isomer ratio no longer changes with further irradiation.

Fig. 4—Typical plot for photoisomerization of *cis*- or *trans*- 1,3-pentadiene in aqueous solution of humic substances (Zepp et al. 1985).

The pseudo-first-order rate constants for this photoisomerization depended strongly on the concentration of dioxygen in the natural waters or humus solutions, indicating that a major fate of the triplets is quenching by dioxygen. Even so, the half-lives observed for the photoisomerization in sunlight were only 2 to 4 hours in air-saturated, colored natural water samples, indicating that triplet sensitized reactions involving energy transfer are likely to be significant for some organic chemicals in aquatic environments.

The relative action spectra for the photoisomerization reaction of pentadiene closely followed that for the dimethylfuran photooxygenation in the ultraviolet region (300–390 nm) (Fig. 3). In the visible region, however, the DMF photoreaction occurred much more efficiently. This difference is attributable to differences in the distribution of triplet state energies available in the chromophores of the humic substances. Pentadiene can accept energy only from chromophores with high energy triplet states (>250 kJ/mol) that absorb in the UV region, whereas dioxygen accepts energy from chromophores with much lower energy triplet states (>94 kJ/mol) including some that absorb visible light. Based on other experiments with humic substances of various origins, it was estimated that up to half of the triplet states that transfer energy to dioxygen had energies in excess of 250 kJ/mol. Fischer and co-workers (1987) reached a similar conclusion based on laser flash photolysis studies of various other humic substances.

Using kinetic results from the pentadiene studies, it was estimated that steady-state concentrations of triplets with energies in excess of 250 kJ/mol are typically about one-fifth as large as singlet oxygen concentrations.

ELECTRON TRANSFER PROCESSES

Most of the research on electron transfer processes involving natural organic matter have focused on electron transfer to dioxygen. This interest mainly was stimulated by research of Cooper and Zika (1983) and Draper and Crosby (1983) who reported that irradiation of various natural water samples leads to production of hydrogen peroxide. The DOM in the water was mainly responsible for photosensitizing formation of the peroxide. Electronically excited states are also known to accept or donate electrons to organic substances (Turro 1978) and it is likely that future research will demonstrate that triplet states of humic substances react directly with certain organic substances via electron transfer reactions.

The light-induced production of hydrogen peroxide in solutions of humic substances and in natural water samples is accelerated in the presence of superoxide dismutase (Cooper and Zika 1983; Petasne and Zika 1987) indicating that the peroxide was formed, at least in part, through the intermediacy of superoxide ions. That the superoxide dismutase diverted an increased part of the superoxide into peroxide formation is in itself an interesting finding.

This result indicated that a large portion of the superoxide was reacting with other species in the water besides itself.

The formation of superoxide could occur in several different ways, including electron transfer from an excited state to dioxygen (Eq. 7), electron photoejection by an excited state (Eq. 11) followed by reduction of dioxygen by the resulting hydrated electrons, as well as other possibilities. Recent results of laser flash studies (Fischer et al. 1987; Power et al. 1987; Zepp et al. 1987) indicate that humic substances do indeed eject electrons when irradiated by UV light. Zepp and co-workers (1987), however, have found that the photoproduction rates of solvated electrons by aquatic DOM and humic substances are too low to account for hydrogen peroxide photoproduction rates observed in sunlight. Solvated electrons, however, react rapidly with biologically refractory chlorinated xenobiotics, and it appears that such reactions may be a significant fate for such toxic chemicals in aquatic environments (Zepp et al. 1987).

Fischer and co-workers (1987) have shown recently that the dication, methylviologen (also known as Paraquat), is rapidly photoreduced by humic substances. This reaction probably involves static photosensitization in which methylviologen sorbed to the humic substances accepts an electron from a nearby chromophore.

FREE RADICAL REACTIONS INVOLVING HUMIC SUBSTANCES

Research on free radicals in natural waters has received a great deal of attention recently. Evidence has emerged that natural organic matter plays a role in the photoproduction and scavenging of free radicals. Some background on this research can be gleaned from the reviews by Zafiriou and co-workers (1984) and Choudhry (1981).

A large portion of the free radicals photoproduced by natural organic matter is likely to be superoxide ions. Their major fate is probably dismutation to form hydrogen peroxide, dioxygen, and hydroxide ions. Recent results obtained by Patasne and Zika (1987), however, indicate that a substantial part of the superoxide formed in coastal seawater participates in other reactions, possibly including reactions with trace metals and certain organic chemicals that are present in aquatic environments. At present, little published information is available on such reactions. Near surface, steady-state concentrations of about 10 nanomols/L have been estimated for superoxide ions during daylight in coastal seawater (Petasne and Zika 1987).

The hydrogen peroxide formed from photochemical interactions of DOM and dioxygen is a more stable oxidant than free radicals. It does, however, decay over a period of a few days in natural waters (Draper and Crosby 1983). The decay potentially could be in part to form the highly reactive free radical, the hydroxyl radical. For example, hydroxyl radicals form through metal-catalyzed decompositions of hydrogen peroxide.

Various methods have been used to detect and quantify free radical intermediates formed from DOM and humic substances. In addition to hydroxyl radicals and superoxide ions, organoperoxyl radicals can form through reactions of superoxide with organic substances or by reactions of organic free radicals with dioxygen. Steady-state irradiations by Mill and co-workers (1980) employed product and kinetic studies of light-initiated oxidations of organic probes to estimate steady-state concentrations of hydroxyl and organoperoxyl radicals in highly colored natural water samples. More recently, Faust and Hoigné (Swiss Federal Institute for Water Resources and Water Pollution Control, Dübendorf-Zürich, unpublished) have concluded that superoxide and organoperoxyl radicals are mainly responsible for the photosensitized oxidation of various alkyl-substituted phenols in solutions of DOM and humic substances. The steady-state concentrations of these organoperoxyl radicals were estimated to be about 3×10^{-11} mol/l, based on their reaction with ascorbate ion. This concentration is about two orders of magnitude lower than that reported earlier by Mill and co-workers (1980) for organoperoxyl radicals.

Other approaches are being used to explore the nature of free radicals formed in natural waters. Zafiriou has used the decay kinetics of the stable radical NO to probe for free radicals in seawater (Woods Hole, unpublished). Blough has been employing electron spin resonance along with various nitroxide scavengers to examine the photoinitiated free radical production from humic acids in water (Woods Hole, unpublished).

In addition to its role in the photoproduction of free radicals, it should be noted that the DOM in freshwater is the primary scavenger of hydroxyl radicals in natural waters (Hoigné and Bader 1979). These scavenging reactions result in the alteration of the humic substances. It is well known, for example, that hydrogen peroxide, which photolyzes to rapidly form hydroxyl radicals when exposed to high intensity short-wavelength light, greatly accelerates the photooxidation of aquatic humic substances (Gjessing 1976). Moreover, in view of the phenolic content of the humic substances themselves, it is possible that superoxide, organoperoxyl radicals, or singlet oxygen may be involved in the light-induced oxidations of DOM. In view of the biologically refractory nature of the colored organic matter in natural waters as demonstrated by Fig. 1, these scavenging reactions, in concert with the direct photoreactions of humic substances that were discussed earlier, could be very important in the cycling of organic carbon in aquatic environments.

PHOTOREACTIVITY OF COMPLEXES FORMED WITH HUMIC SUBSTANCES

Limited research has been conducted on light-induced processes involving trace metals and humic substances. This research area has been briefly reviewed by Zafiriou and co-workers (1984). A much more extensive review by Waite (1986) of the photoreactions of metal oxides and various natural organic ligands, including humic substances, has recently appeared. One significant conclusion from these studies is that photochemical processes involving natural organic matter play an important role in redox cycling of iron and manganese in the photic zones of natural water bodies.

Interactions between humic substances and metals also result in static quenching of excited state processes of the humic substances. For example, complexation of trace metals by humic substances has been shown by laser flash studies to quench their fluorescence (Power et al. 1986) and the absorption attributable to triplets (Frimmel et al. 1987)

Less is known about photoreactions of organic chemicals "bound" to humic substances. Zepp and co-workers (1985) provided evidence that static photosensitization could not account for the results that were described earlier for pentadiene and dimethylfuran. On the other hand, Gauthier and co-workers (1986) provide lead references to studies that indicate that extremely hydrophobic pollutants (e.g. polycyclic aromatic hydrocarbons, DDT, mirex) have a strong tendency to associate with the organic matter in water bodies. These chemicals are up to six orders of magnitude less water soluble than DMF and pentadiene. The results of Gauthier and co-workers (1986) indicate that the fluorescence of polycyclic aromatic hydrocarbons is quenched by their association with humic substances. This static quenching possibly indicates that the photochemistry also would be quenched.

On the other hand, recent unpublished results of Hassett (State University of New York, Syracuse) indicate that association of the highly chlorinated pollutant, mirex, with the DOM in Lake Ontario water leads to a large enhancement in the rate of photoproductive dechlorination of this persistent pollutant. This reaction perhaps represents another example of static photosensitization involving electron transfer. Further research in this area will no doubt be forthcoming in the near future.

Acknowledgements. The author gratefully acknowledges technical comments on this paper by A. Braun, L. Azarraga, N. Blough, M. Ewald, F. Frimmel, J. Hoigné, N. Loux, O. Zafiriou, and R. Zika.

REFERENCES

Baker, K.S., and Smith, R.C. 1982. Bio-optical classification and model of natural waters. 2. *Limnol. Oceanogr.* **27**: 500–509.

Chapman, O.L. 1967. Organic Photochemistry, vol. 1, p. 127. New York: Dekker.

Choudhry, G.G. 1981. Humic substances. Part II: Photophysical, photochemical, and free radical characteristics. Toxicol. Env. Chem. **4**: 261–295.

Cooper, W.J., and Zika, R.G. 1983. Photochemical formation of hydrogen peroxide in surface and ground waters exposed to sunlight. Science **220**: 711–712.

Donard, O.F.X.; Belin, C.; and Ewald, M. 1987. Corrected fluorescence excitation spectra of fulvic acids. Comparison with the UV/visible absorption spectra. Sci. Total. Env. **62**: 157–161.

Draper, W.M., and Crosby, D.G. 1983. The photochemical generation of hydrogen peroxide in natural waters. Arch. Env. Contam. Toxicol. **12**: 121–126.

Fischer, A.M.; Winterle, J.S.; and Mill, T. 1987. Primary photochemical processes in photolysis mediated by humic substances. In: Photochemistry of Environmental Aquatic Systems, ed. R.G. Zika and W.J. Cooper, ACS Symposium Series 327, pp. 141-156. Washington, D.C.: American Chemical Society.

Frimmel, F., and Bauer, H. 1987. Influence of photochemical reactions on the optical properties of aquatic humic substances gained from fall leaves. Sci. Total. Env. **62**: 139–148.

Frimmel, F.H.; Bauer, H.; Putzien, J.; Murasecco, P.; and Braun, A.M. 1987. Laser flash photolysis of dissolved aquatic humic material and the sensitized production of singlet oxygen. Env. Sci. Tech., in press.

Gauthier, T.D.; Shane, E.C.; Guerin, W.F.; Seitz, W.R.; and Grant, C.L. 1986. Fluorescence quenching method for determining equilibrium constants for polycyclic aromatic hydrocarbons binding to dissolved humic materials. Env. Sci. Tech. **20**: 1162–1166.

Gjessing, E.T. 1976. Physical and Chemical Characteristics of Aquatic Humus. Ann Arbor: Ann Arbor Sci. Publishers Inc.

Gohre, K.; Scholl, R.; and Miller, G.C. 1986. Singlet oxygen reactions on irradiated soil surfaces. Env. Sci. Tech. **20**: 934–938.

Gorman, A.A., and Rodgers, M.A.J. 1986. The quenching of aromatic ketone triplets by oxygen: competing singlet oxygen and biradical formation? J. Am. Chem. Soc. **108**: 5074–5078.

Haag, W.R., and Hoigné, J. 1986. Singlet oxygen in surface waters. Part III: Photochemical formation and steady state concentrations in various types of waters. Env. Sci. Tech. **20**: 341–348.

Haag, W.R.; Hoigné, J.; Gassmann, E.; and Braun, A.M. 1984. Singlet oxygen in surface waters. Part II: Quantum yields of its production. Chemosphere **13**: 641–650.

Harvey, G.R.; Boran, D.A.; Chesal, L.A.; and Tokar, J.M. 1983. The structure of marine fulvic and humic acids. Marine Chem. **12**: 119–132.

Hoigné, J., and Bader, H. 1979. Ozonation of water. Oxidation competition values (for OH – radical reactions) of different types of Swiss natural waters. Ozone Sci. Eng. **1**: 357–372.

Højerslev, N.K. 1982. Yellow substance in the sea. In: The Role of Solar Ultraviolet Radiation in Marine Ecosystems, ed. J. Calkins, pp. 263–281. New York: Plenum Press.

Hutchinson, G.E. 1975. A Treatise on Limnology, vol I, part 2. Chemistry of Lakes. pp. 878–902. New York: Wiley.

Il'in, N.P., and Orlov, D.S. 1979. The principle and methods of photochemical studies of humic substances. Agrokhimiya **3**: 120–129.

Kouassi, A.M. 1986. Light induced alteration of the photophysical properties of dissolved organic matter in seawater. M.S. Thesis, Univ. of Miami, Coral Gables, FL.

Kouassi, M.; Zika, R.G.; Plane, J.M.C.; and Gidel, L. 1986. Photochemical modelling of marine humus fluorescence in the ocean. *Trans. Am. Geophys. U.* **66**: 1266.

Malcolm, R.L., and McCarthy. P. 1986. Limitations in the use of commercial humic acids in water and soil research. *Env. Sci. Tech.* **20**: 904–911.

Mill, T.; Hendry, D.G.; and Richardson, H. 1980. Free radical oxidants in natural waters. *Science* **207**: 886–887.

Milne, P.J.; Odum, D.S.; and Zika, R.G. 1987. Time-resolved fluorescence measurements on dissolved marine organic matter. In: Photochemistry of Environmental Aquatic Systems, ed. R.G. Zika and W.J. Cooper, ACS Symposium Series 327, pp. 132–140. Washington, D.C.: American Chemical Society.

Mopper, K., and Zika, R.G. 1987. Natural photosensitizers in sea water. In: Photochemistry of Environmental Aquatic Systems, ed. R.G. Zika and W.J. Cooper, ACS Symposium Series 327, pp. 174–190. Washington, D.C.: American Chemical Society.

Ononye, A.I.; McIntosh, A.R.; and Bolton, J.R. 1986. Mechanism of the photochemistry of *p*-benzoquinone in aqueous solutions. Spin trapping and flash photolysis electron paramagnetic resonance studies. *J. Phys. Chem.* **90**: 6266–6274.

Petasne, R.G., and Zika, R.G. 1987. Fate of superoxide in coastal sea water. *Nature* **325**: 516–518.

Power, J.F.; LeSage, R.; Sharma, D.K.; and Langford, C.H. 1986. Fluorescence lifetimes of the well-characterized humic substance, Armadale fulvic acid. *Env. Tech. Lett.* **7**: 425–430.

Power, J.F.; LeSage, R.; Sharma, D.K.; Langford, C.H.; Bonneau, R.; and Joussot-Dubien, J. 1987. Laser flash photolytic studies of a well-characterized soil humic substance. In: Photochemistry of Environmental Aquatic Systems, ed. R.G. Zika and W.J. Cooper, ACS Symposium Series 327, pp. 157–173. Washington, D.C.: American Chemical Society.

Thurman, E.M. 1985. Organic Geochemistry of Natural Waters. Boston: Martinus Nijhoff/Dr.W. Junk Publishers.

Turro, N.J. 1978. Modern Molecular Photochemistry. Menlo Park, CA: Benjamin Cummings Publishing Co.

Usui, Y., and Kamagawa, K. 1974. A standard system to determine the quantum yield of singlet oxygen formation in aqueous solution. *Photochem. Photobiol.* **19**: 245–247.

Waite, T.D. 1986. Photoredox chemistry of colloidal metal oxides. In: Surface Processes in Aqueous Geochemistry, eds. J.A. Davis and K.F. Hays, ACS Symposium Series 323. Washington, D.C.: American Chemical Society.

Whipple, G.C. 1914. The Microscopy of Drinking Water. New York: Wiley.

Zafiriou, O.C.; Joussot-Dubien, J.; Zepp, R.G.; and Zika, R.G. 1984. Photochemistry of natural waters. *Env. Sci. Tech.* **18**: 358A–371A.

Zepp, R.G.; Baughman, G.L.; and Schlotzhauer, P.F. 1981. Comparison of photochemical behavior of various humic substances in water: photosensitized oxygenations. *Chemosphere* **10**: 119–126.

Zepp, R.G.; Braun, A.M.; Hoigné, J.; and Leenheer, J.A. 1987. Photoproduction of hydrated electrons from natural organic solutes in aquatic environments. *Env. Sci. Tech.*, in press.

Zepp, R.G.; Schlotzhauer, P.F.; and Sink, R.M. 1985. Photosensitized transformations involving electronic energy transfer in natural waters: role of humic substances. *Env. Sci Tech.* **19**: 48–55.

Zepp, R.G.; Wolfe, N.L.; Baughman, G.L.; and Hollis, R.C. 1977. Singlet oxygen in natural waters. *Nature* **267**: 421–423.

Humic Substances and Their Role in the Environment
eds. F.H. Frimmel and R.F. Christman, pp. 215–243
John Wiley & Sons Limited
© S. Bernhard, Dahlem Konferenzen, 1988.

Sedimentology of Organic Matter

A.Y. Huc

*Institut Français du Pétrole
92506 Rueil, France*

Abstract. Organic matter is the precursor of oil and gas generated in sedimentary basins. As such, the factors governing the occurrence and formation of organic-rich beds deserve to be understood in order to provide conceptual guidelines for source rocks evaluation in basins under petroleum exploration.

Organic sedimentology is concerned with the genesis of organic-bearing sediments and with the factors governing the facies distribution in sedimentary basins. The main factors controlling the accumulation of organic matter and the patterns of organic facies distribution include the production of biomass, degradation, and transport processes.

The preservation of organic matter in the geological record is the result of a minute escape from the carbon cycle in the biosphere. However, dramatic changes of the absolute amount of this escape occur on a worldwide scale during earth history.

INTRODUCTION

The very necessity of organic sedimentology is based on the following facts:

1) Organic matter is a minor but normal component of sediments.
2) The organic content of sediments is highly variable, ranging from less than 0.1% in oceanic red clays up to almost 100% in certain humic or algal coals.
3) The organic matter in sediments consists of several distinguishable types according to their hydrocarbons potentials.
4) The distribution of organic facies is obviously not random.

Organic sedimentology is basically concerned with the genesis of organic-bearing sediments and with the factors governing the organic facies distribution in sedimentary basins. The practical input of such a knowledge is of paramount importance because the quantity and quality of organic matter occurring in sediments are obviously key factors controlling the amount of

215

hydrocarbons generated in a given basin. Moreover, as stated by current theoretical considerations, expulsion of hydrocarbons out of source rocks is believed to proceed as a separate phase (Durand 1983). As a consequence the efficiency of migration is dependent on the permeability of the system to hydrocarbons. According to Darcy's law, extended to polyphasic flow, there is competition between hydrocarbons and water in terms of relative permeability, and the hydrocarbons are expelled only when the hydrocarbons saturation will be above a minimum threshold. In other words, the absolute quantity of hydrocarbons which is delivered in the pore system as maturation progresses (which is locally controlled by the quantity and quality of kerogen and by the thermal history of the sediment) is determinant for the initiation of expulsion mechanisms and, thus, for the amount of hydrocarbons which are able to circulate in the reservoir rocks, and eventually be trapped. Consequently, a good knowledge of the source rocks, including their actual presence, stratigraphic placement, specific location in sedimentological sequences, lateral extension, geometry at the scale of the basin, and organic facies variations (richness and quality), is significant information to be considered to reduce geologic risk in petroleum exploration.

To achieve predictions in terms of source rocks, explorationists must currently rely on data obtained on samples from outcrops and wells, which are not always located in relevant areas as far as source rocks are concerned (drilling operations are aimed to find oil pools, not to study source rocks). Moreover, samples from exploration wells are essentially cuttings with their inherent inconveniences including caving and mud contamination, imprecise sampling depth, etc.; cores being usually taken in reservoirs but rarely in source beds.

Subsequently, in order to supplement source rock surveys based on analytical data banks, conceptual tools designed to predict organic facies distributions are needed. These concepts have to be derived from the basic knowledge of the factors controlling the organic matter sedimentology. For several years a great deal of effort has been devoted to this topic, including investigations in modern sedimentary environments and in basins under petroleum exploration, in order to decipher the critical factors involved in the deposition of source beds and to try to model them.

The qualitative and quantitative status of organic matter in a given sediment (the organic facies as defined by Demaison and Moore (1980)) is basically the result of the combined influence of biomass productivity, biochemical degradation of organic matter, and depositional processes.

PRODUCTION OF BIOMASS

Figure 1, derived from a large compilation/interpolation by Romankevitch (1977), shows the distribution of the concentration of organic carbon in modern sediments. The very first observation is that the organic-rich areas are located on the continental margins and in the epicontinental seas; in other words, on the outward bounds of the emerged landmasses. Such a pattern is obviously a reflection of a dependence on the proximity of organic sources.

The basis for the production of primary organic matter is photosynthesis which occurs on land and in the euphotic layer of the seas (upper 100 m). A small portion (ca. 0.5%) of the organic matter which is produced on the present-day land surface (productivity on land is estimated at $100–180 \times 10^9$ tons/year) escapes the continental biological cycle to ultimately make its way into the seas where a part of it eventually accumulates in coastal environments. Consequently, significant accumulations of terrestrial organic matter are more likely to occur at the outlets of rivers which are the main conveyors in this process. In this respect, and according to several present-day examples (Mahakam, Niger deltas), deltaic and related environments are good examples of production/accumulation of terrestrial organic matter (organic shales, coals) on continental margins.

Moreover, as depicted in Fig. 2 (after Degens and Mopper 1976), most of the areas of high marine organic productivity are also located in coastal environments. In the euphotic zone, the site of aquatic photosynthesis, primary productivity is mainly controlled by the local availability of nutrients such as phosphates and nitrates. Phytoplankton growth leads to a rapid impoverishment of nutrients in superficial waters. As a result, open oceans are generally poor in primary productivity and are considered to be biological deserts. Subseqently, high productivity occurs only in specific areas where these nutrients can be replenished at a sufficient rate. Such a situation is favored in inshore and nearshore regions where rivers can supply large amounts of nutrients originating from continental runoff and where upwelling of deep ocean water can introduce significant quantities of nutrients in the photic zone. Upwelling currents occur when surface waters are driven offshore by winds, allowing subsurface waters rich in nutrients to well up and take their place, stimulating high productivity.

Consequently, 55% of the present-day marine productivity, which is estimated at $20–60 \times 10^9$ tons/year, occurs in estuaries, coastal lagoons, reefs, and continental shelf areas. Lake surface represents only 1% of the ocean surface but their primary productivity, usually activated by the large supply of nutrients from the surrounding landmasses, is up to 100 times higher per surface unit than ocean productivity. This situation is particularly favorable for organic matter accumulation, and many examples of lakes with especially rich organic sediments are reported in recent environments (lake Tanganyika, lake Kivu, etc.) as well as in the geological record (Green River shales of Eocene age in Utah, lake Cadell of lower Carboniferous age in Scotland, the Dongying depression of Oligocene age in China, etc.).

218

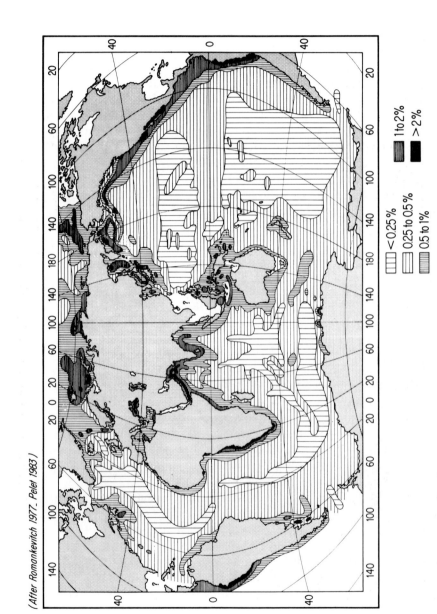

(After Romankevich 1977. Pelet 1983)

<0.25%

0.25 to 0.5%

0.5 to 1%

1 to 2%

>2%

Fig. 1—Distribution of organic carbon in surficial sediments (% TOC).

(After Degens et al 1976, Pelet 1983)

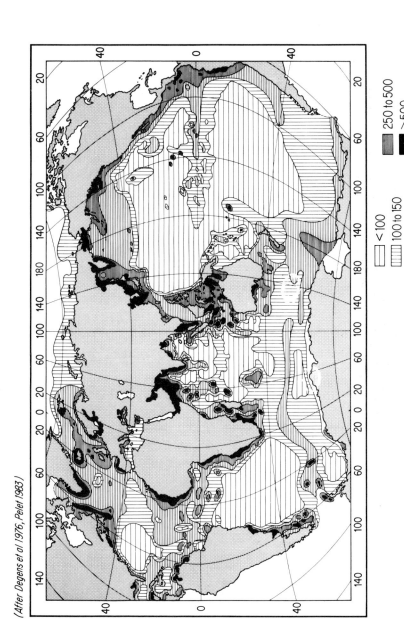

Fig. 2—Distribution of primary productivity in the world ocean (mg C/m² × day).

Overall, there is a reasonable relationship between organic-rich sediments and high biomass productivity areas, which in turn are dependent on the climate and the geographic setting. Schematically, high productivity on land is encountered in regions of high rainfall (Fig. 3). This situation induces intensive runoff and a large supply of terrestrial organic matter and nutrients to the adjoining seas and lakes. In the marine realm, the most extensive zone of high productivity is related to regions of upwelling currents. Consequently, to a large degree, productivity can be related to atmospheric circulation patterns and landmasses configuration. This concept has led several authors (Parrish 1982, Parrish and Curtis 1982, Barron 1985, Scotese and Summerhayes 1986) to apply numerical climate models, based on the fundamental physical laws driving atmospheric and oceanic circulation, for times of different continental configurations during earth history. Using these computer simulations of paleoclimates, several maps have been constructed to tentatively localize upwelling zones and high rainfall regions during different geological periods (Fig. 4). These predictions have been compared to the actual distributions of organic-rich rocks and coals of corresponding age. Results are far from being entirely conclusive but several good correspondences support their approach to the source-rock problem. If such predictions cannot be taken as granted, they should be regarded as "an additional indicator of increased probability of a source rock location in conjunction with the entire suite of more traditional geologic data" (Barron 1985).

DEGRADATION OF ORGANIC MATTER

Living tissues are composed of an assemblage of biomolecules which are thermodynamically unstable. As soon as these biomolecules are no longer involved in living processes (i.e., when they are secreted or excreted, or after the death of organisms) they tend to lose their integrity and can be ultimately transformed into simple stable components such as CO_2, H_2O, CH_4, NH_4^+, etc. This degradation can rely on physicochemical processes (oxidation, photolysis, thermolysis, etc.) but is by far mainly biologically mediated.

Organic matter is actually a basic source of nutrients and energy for heterotrophic living organisms including consumers (zooplankton, nekton, zoobenthos) and decomposers (microbial communities).

The processes and end-products of organic matter decomposition are to a large extent controlled by the availability of electron acceptors. The presence of adequate oxygen concentration (atmospheric or dissolved in water) provides a suitable living medium for organisms ranging from aerobic microbes to higher animals (grazers, feeders, burrowers etc.). In such a situation, the overall decomposition process is oxidation using oxygen as an electron acceptor.

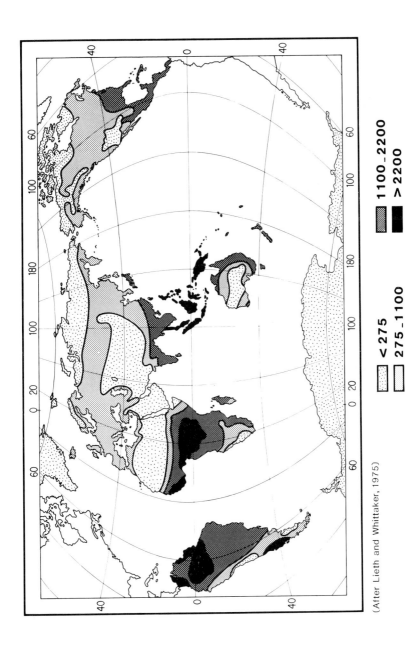

(After Lieth and Whittaker, 1975)

< 275

275_1100

1100_2200

>2200

Fig. 3—Distribution of primary productivity on land (mg C/m² × day).

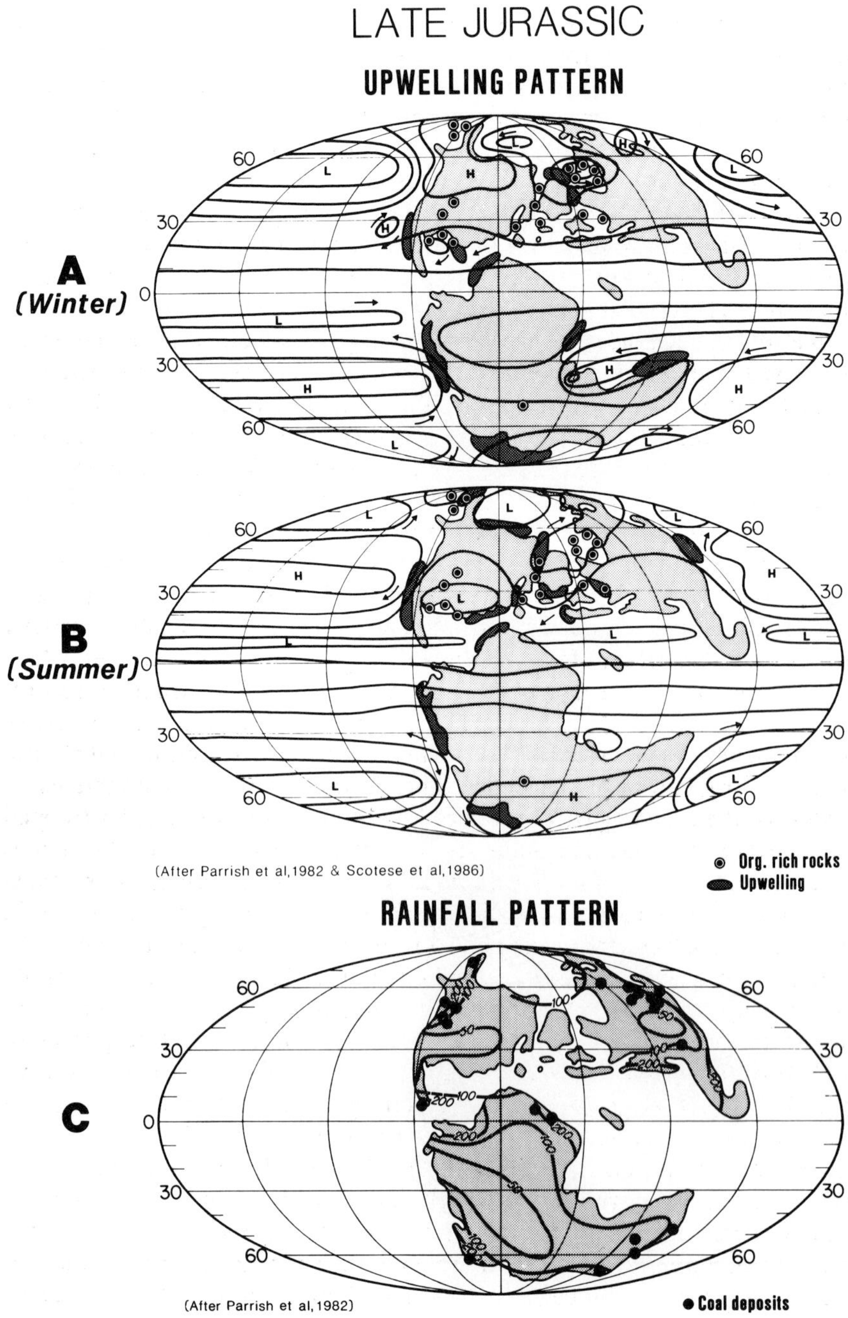

LATE JURASSIC

UPWELLING PATTERN

A
(Winter)

B
(Summer)

(After Parrish et al,1982 & Scotese et al,1986)

⦿ Org. rich rocks
⬤ Upwelling

RAINFALL PATTERN

C

(After Parrish et al,1982)

● Coal deposits

Fig. 4–Computerized upwelling and rainfall patterns during late Jarassic; relationship with organic-rich rocks and coal deposits.

If molecular oxygen is no longer available no organisms higher than bacteria can survive; nitrates followed by sulfates are used as an oxygen source by anaerobic microorganisms in order to oxidize organic matter. Ultimately, when the medium is devoided of oxidants (O_2, NO_3, SO_4), fermentative degradation proceeds using the organic matter itself as an electron acceptor (via, CO_2 and acetate reduction).

As far as biological factors are concerned, it is apparent that the efficiency of aerobic decomposition is likely to be enhanced by the mechanical and enzymatic breakdown of tissues during feeding and digestion by higher organisms. These processes are lacking in anoxic environments.

Due to these degradative processes the organic matter which is preserved in the sedimentary record represents only a small fraction (few tenths of percent) of the primary production. This organic matter, which ultimately escapes the biological cycle, is composed of the end-products of the decomposition of the various organisms involved in the trophic chain including producers (i.e., phytoplankton), consumers (i.e., zooplankton), and decomposers (microbial communities). Biological degradation is thus one of the main controls on the quality and quantity of organic matter deposited in sedimentary basins. The degradative biological processes begin to proceed as soon as organic components are no longer involved in metabolic activity and persist throughout the subsequent sedimentological processes which drive the organics toward their final burial as a part of the sediment record. As a consequence, the extent of organic alteration depends on the successive environments encountered by the organic matter throughout its sedimentological history and on its residence times in these various environments.

Most of the organic matter produced on the land surface remains trapped in soils where humification processes take place; a small part of this humus can eventually be transported toward sedimentary basins (oceans, seas, lakes) by drainage mechanisms. Due to the extent of the biological reworking in the soils themselves, and during the transportation from the catchment, the organic material which is usually supplied to the sedimentary basins is heavily altered (impoverished in hydrogen and nitrogen) and tends to be refractory to further biological processing. Thus, usually, the chemical properties of land-derived organic material reflect its lignocellulosic source and its extended alteration (mainly oxidative) during transportation and results in a hydrogen-depleted organic material in the sediment record.

Peat bogs are a significant exception to this rule. In this case, the organic matter accumulates without transportation, and the high concentration of organic components in the medium can result in a high acidity and high concentration of biostatics, such as phenolic compounds derived from lignin, which might hamper biological activity and limit the degradation of organic matter. This is well illustrated by the near perfect preservation of several hundred year old wooden artifacts, clothing, and even human corpses discovered in certain northern European peats.

In the aquatic environment, the fresh organics, which are supplied below the base of the euphotic zone as a result of primary production, are highly reactive and are likely to be intensively degraded by heterotrophic organisms. In fact, simple mass balance calculations (organic production versus organic material actually incorporated into the underlying sediments) suggest that degradation of organic detritus proceeds to a great extent in the water column. In the last few years, sediment trap experiments have refined our understanding of the organic carbon flux in the oceans and have shown the quasi-exponential destruction of organic matter as a function of water depth/ residence time (Fig. 5 after Suess 1980).

The low density, pure organic detritus (1 to 1.7 g/ml) are unlikely to play a significant role in vertical mass flux because their high residence time in the water column, due to their buoyancy, ultimately favors their degradation. According to Stoke's law, small organic or organomineral particles have a

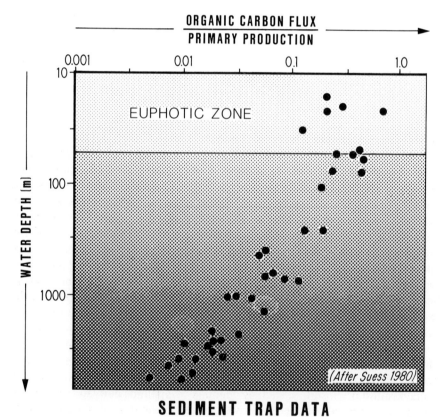

Fig. 5—Degradation of organic material in the oceanic water column according to sediment trap data.

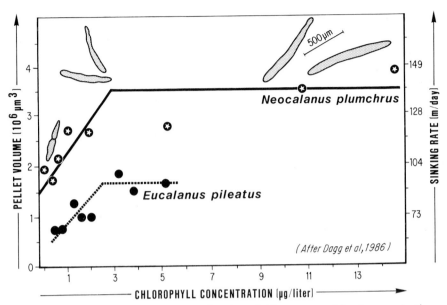

Fig. 6—Relationship between fecal pellet size, sinking rate, and food concentration.

low sinking rate and hence a high residence time in the water column. However, repacking of organic detritus and of small particles by physico-chemical (i.e., flocculation) and biological processes produces large organo-mineral particles (fecal pellets, marine snow, aggregates) with rapid settling rates which can act as vehicles for the organic matter. One can stress that the biomass itself is responsible for at least a part of its own preservation during transportation through the water column by providing pertinent vehicles: the fecal pellets, which result from the feeding processes of the fauna and presumably the marine snow made of aggregates which are sup-posed to be mainly mediated by net-like muco-polysaccharide biopolymers.

In this respect, interesting observations have been performed by Dagg and Walser (1986) who reported that properties of fecal pellets are related to the level of food availability (primary productivity). Fecal pellets produced by zooplankton under conditions of low food availability are small and loosely packed and exhibit a low sinking rate; thus they are less likely to sink out of the euphotic zone than pellets produced under conditions of higher food concentration (high ˙primary productivity), which are larger, more densely packed and have a significantly higher sinking rate (Fig. 6). Consequently, high primary productivity not only provides a large amount of organic matter but promotes a higher preservation and higher transport efficiency toward the underlying sediments.

In addition to the recycling in the water column, organic matter is likely to experience further biological alteration during its postdepositional history. In this respect the difference between oxic and anoxic depositional environments has been clearly demonstrated by Demaison and Moore (1980).

In oxygenated bottom environments, a significant percentage of the organic matter is consumed by benthic fauna on the seafloor and by burrowing organisms in the surface sediments (Pelet 1984). Furthermore, the burrowing activity maintains a circulation of water replenishing electron acceptors (dissolved oxygen, sulfates) into the sediment, fuelling bacterial oxidative degradation of organic matter.

In contrast, the anoxic environment is toxic to macro- and meiobenthic fauna including burrowers and results in undisturbed laminated sediments in which water circulation is strongly limited (Fig. 7). In such an environment organic preservation is enhanced by the lack of benthic animal scavengers and by the limited supply of electron acceptors into the sediment.

In oxic environments the maintenance of living activity of bottom-dwelling animals results in a mixing of the upper layer of sediments (bioturbation) which significantly increases the time of exposure of organic matter to decomposition processes. However, as benthic organisms are active only in the 10–15 cm top layer of sediments it can be expected that rapid sedimentation and burial can enhance organic preservation owing to a rapid removal of organic material from the zone of of bioturbation. For instance, in the sediments underlying the upwelling area located off the coast of Angola (DSDP site 532), the high productivity promotes a high sedimentation rate of biogenic siliceous and carbonates sediments (50 cm/100 years since the Miocene with values as high as 200 cm/1000 years during the Pleistocene) resulting in a high organic content (3–8% TOC) together with clear evidence of burrowing structures.

From a large compilation of data Muller and Suess (1979) and Stein (1986) have postulated that a positive relationship exists between sedimentation rate and organic content in surface sediments deposited in oxygenated seawater environments (Fig. 8). However, if a high sedimentation rate seems to be a favorable factor for organic preservation, too high a sedimentation rate might result in a decrease of the relative organic content, especially in detrital situations, owing to dilution effect of the organic input (Ibach 1982; Ten Haven 1986).

On the other hand, in anoxic environments the absence of a near-surface zone of intense organic degradation limits the influence of the sedimentation rate on organic preservation. As a matter of fact, the data presented by Stein (1986) suggest an absence of positive relationship between sedimentation rate and sedimentary organic content in anoxic environments (Fig. 8).

These observations, however, should be carefully considered because, as pointed out by Pelet (1987), possible covarying factors such as water depth

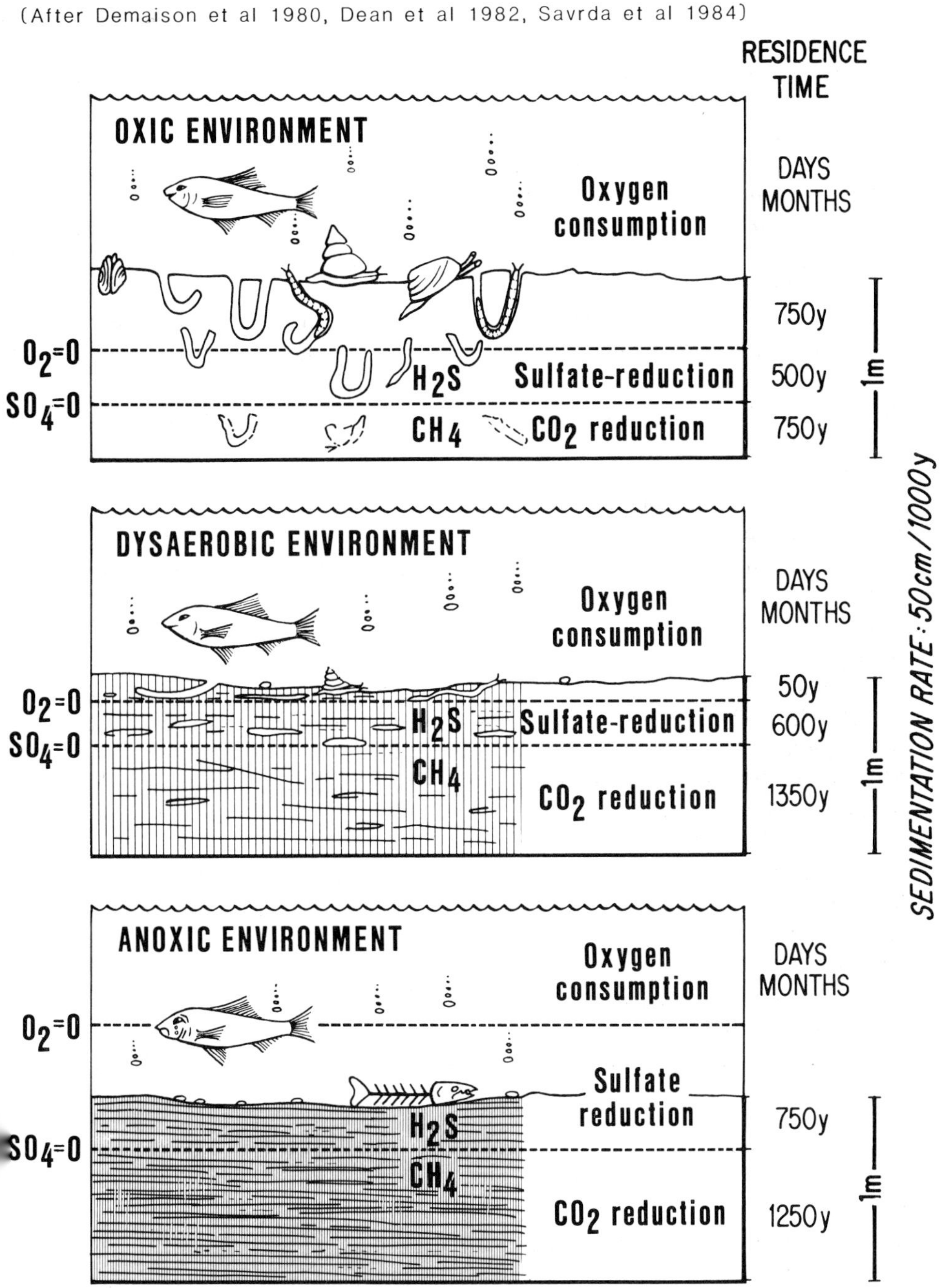

(After Demaison et al 1980, Dean et al 1982, Savrda et al 1984)

Fig. 7—Oxic, dysaerobic, and anoxic depositional environments.

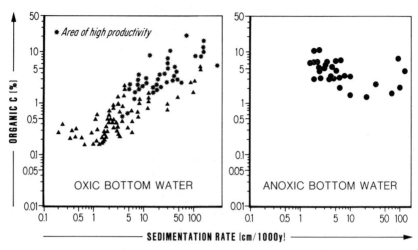

RECENT AND QUATERNARY SEDIMENTS

(After a compilation by Stein 1986 & data from Pelet 1983)

Fig. 8—Relationship between organic carbon content (%) and sedimentation rate in recent and quaternary sediments.

and distance from the shore, which are relevant factors in terms of organic sedimentation (degradation, organic input...), are likely to tangle the sedimentation rates influence. In such instance, an apparent linear correlation does not mean a direct link between organic content and sedimentation rate. Moreover, bilogarithmic representation which is used in these diagrams enhances the relationship between parameters which otherwise would have been rather vague.

Logical considerations lead to the assumption that degradation processes are not random but affect preferentially the more labile and energy-bearing fractions of the organic matter. As a result it can be supposed that the extent of organic alteration affects not only the quantity of preserved sedimentary organic matter but also its quality and particularly its petroleum potential. Pratt (1984) has provided arguments supporting this idea by reporting a convincing relationship between quantity and quality of organic matter and the extent of bioturbation in cretaceous sediments from the North American Western Interior Seaway. In this example, the higher the biodegradation is, the lower the organic content and hydrogen index of the organic matter are (the hydrogen index is defined as the quantity of hydrocarbons released by ROCK-EVAL pyrolysis and standardized to organic carbon). On the other hand, when bioturbation is lacking the sediments are finely laminated and are characterized by high concentrations of organic matter exhibiting high values of the hydrogen index (Fig. 9).

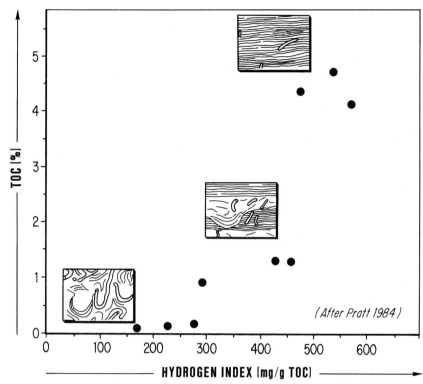

Fig. 9—Relationship between organic matter and degree of bioturbation in the Western interior seaway, U.S.A. (Middle Cretaceous).

According to the previous remarks it is obvious that the biochemical degradation of organic matter is strongly dependent on the conditions of environment and more specifically on the presence of oxygen. Anoxic conditions develop at a given location when the rate of oxygen consumption by respiratory processes involved in organic degradation outbalances the rate of oxygen replacement. In aquatic environments two fundamental controls govern such a situation: a) the organic input which is responsible for the level of oxygen demand, and b) the circulation of oxygen-rich water masses which is responsible for the level of oxygen supply.

The primary source of oxygen-rich water is the surface layer into which oxygen is supplied by exchange through the water/atmosphere interface and by photosynthesis in the euphotic zone. Hydrodynamic movements are responsible for driving such oxygenated water masses throughout aquatic basins. These movements are a consequence of the relative densities of water masses and atmospheric circulations. In polar regions, for instance, cooled oxygen-rich surface waters with high density (due to its low temperature) tend to sink to the bottom and circulate along the ocean floor toward lower latitude regions where they eventually rise back to the surface in upwelling areas. Such a global cell circulation is responsible for the nowadays good ventilation of the bottom of the world ocean, preventing organic preservation in deep oceanic basins. The low organic content of the biogenic opaline sediments deposited in open ocean beneath the upwellings located along the equator exemplifies this lack of organic preservation in well ventilated modern deep oceanic basins (Figs. 1, 2, 10).

Development of anoxic conditions are favored by water stratifications and topographic features such as sills or seafloor depressions interfering with oxygen circulation patterns. Stratification is initiated by density differences of water masses: cold waters and salty waters are denser than warm waters and fresh waters. Consequently, occurrences of stratification can be explained as a result of various situations depicted in Fig. 11.

An example of local topographic control responsible for the development of anoxia is provided in Kaoe bay offshore Java, where a deep basin partially surrounded by emergent lands is isolated from the global oceanic circulation by the occurrence of a sill. Organic matter production and preservation results in sediments with high organic carbon content (2–3% TOC) (Fig. 12, after Debyser and Deroo 1969).

It should be emphasized, however, that these confining conditions are only factors favoring the development of anoxia and organic preservation but that they do not automatically imply the accumulation of organic matter in bottom sediments. As an example, brines seeping out of the sediments into the Red Sea result, when occurring in seafloor depressions, in local strong water stratification. As a consequence of this situation and of the high temperature of the brines, the dissolved oxygen content of bottom water is very low in these depressions. Nevertheless, underlying sediments are organic-poor due to the lack of organic matter input (low surface productivity).

Through geological time, larger scale topographic controls were probably markedly influenced by landmass configurations. A good example is provided by the deposition of organic-rich sediments in the narrow, restricted, early Albian proto-Atlantic basin at the beginning of the South Atlantic opening (Fig. 13 after Herbin et al. 1986)

Together with paleotopography, sea level change has been advocated as a major control on organic matter preservation at the world scale (Tissot 1979): transgressive periods correspond to extensive landlocked epicontinental seas over continental depressions away from global oceanic circulation. This situation is likely to result in insufficient oxygen replenishment to fulfill the oxygen demand required by the oxidative consumption of the organic matter showering into these presumably highly productive seas, since epicontinental marine areas are usually adequately supplied with nutrients by surrounding landmasses.

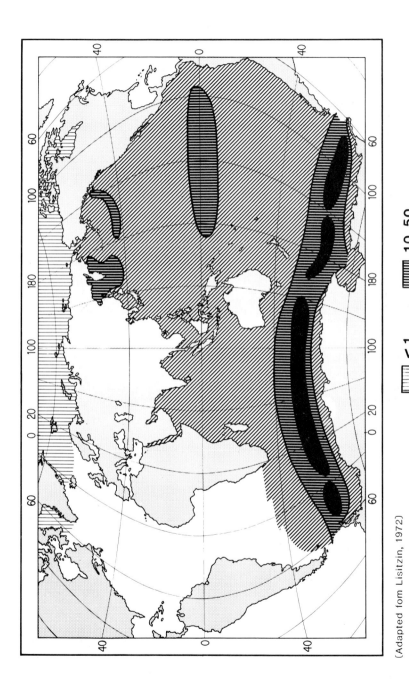

(Adapted fom Lisitzin, 1972)

Fig. 10—Distribution of amorphous silica in surface sediment layer (% dry sediment).

<1   10–50
1–10 >50

STRATIFICATION MODELS

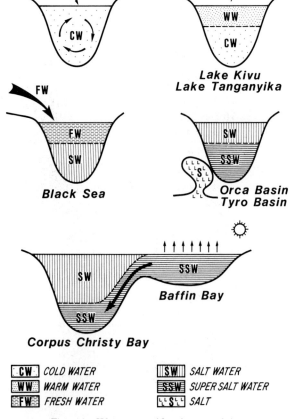

Fig. 11—Water stratification models.

SEDIMENTATION PROCESSES

Most of the studies devoted to the deposition of organic-rich sediments emphasize an approach based on global basin conditions including productivity and preservation, but with few exceptions (Arthur et al. 1984; Stanley 1986) they neglect the fact that organic matter is composed of physical particles, as clays or sands are, and that its lateral distribution inside a given basin is governed by sediment transport rules. This evidence is clearly illustrated when one considers the well-studied Black Sea: the Black Sea is a modern example of a deep (2200 m) anoxic silled basin. A look at the maps depicted in Fig. 14 (after Shimkus and Trimonis (1974)) shows that

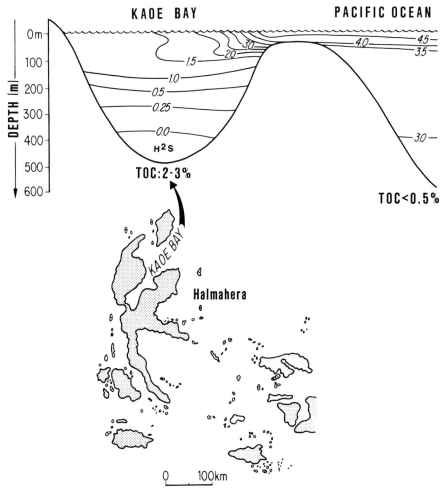

Fig. 12—Distribution of dissolved oxygen in Kaoe Bay (cm³/l) (from Debyser et al. 1969).

the distribution of the sedimentary organic matter is independent of the distribution of the organic productivity in surface waters and of the distribution of anoxic water masses which overlie the entire bottom sediment surface below the halocline. On the other hand, a close relationship between the organic matter content and the mineral grain size distribution is remarkable. In the Black Sea the finer sediments settle in the central areas and especially in the hydrodynamically quiescent areas setting in the nodal zones of cyclonic currents. This preferential concentration of organic matter together with the finer sediments is probably the result of hydraulic equivalent

234

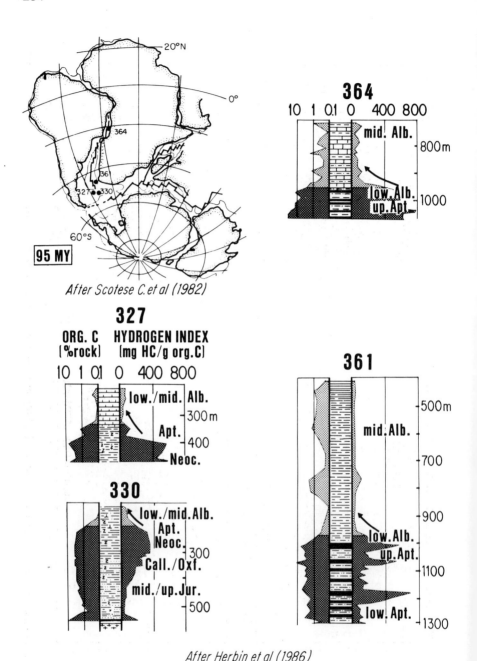

Fig. 13—Organic content in the Proto-Atlantic sediments.

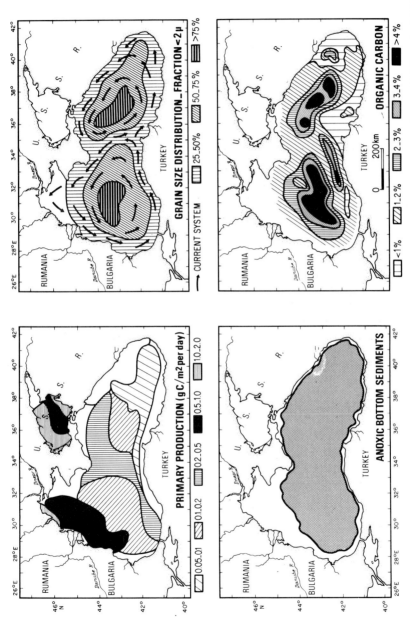

Fig. 14—Distribution of organic carbon in the bottom sediments of the Black Sea, after Shimkus and Trimonis (1974).

effects. As a matter of fact, densities ranging from 1.04 to 1.72 g/ml have been reported by Nwachukwu and Baker (1985) for organic particles in recent sediments. Such a low density, added to the readiness of organic matter to carry on physicochemical associations with clay minerals, favors the deposition of organic matter in depocenters of lowest energy along with hydraulically equivalent small grain size mineral particles.

Besides pelagic and hemipelagic processes, however, sediments can be redeposited from a primary settling environment by mass-gravity transport. Turbidity currents and related gravity flows move sediments, including their organic content, down relatively steep slopes and eventually redistribute them in areas which are not predicted by simple organic depositional models such as anoxia and hydraulic energy level. This situation results in an apparently complex organic facies distribution as exemplified by the organic maps constructed for Lake Tanganyika (Huc, Vandenbroucke et al. 1987). This lake, which is lying in the East African intracontinental rift system, is characterized by a steep topography pattern controlled by faults and tilted blocks. The hydrology and the climate of the Tanganyika basin ensure that thermal stratification of waters occurs. These factors, combined with a high planktonic productivity, result in substantial organic accumulation in bottom sediments (Degens et al. 1971). However, the sediment cover of this lake exhibits a complex organic facies distribution obviously due to the development of gravity flows which originate from oxic shallow areas poor in organic matter and which descend to the deep floor of the lake to interlayer with organic-rich pelagic/hemipelagic muds deposited under anoxic conditions. The shallow water origin of the organic matter in these gravity flow deposits is supported by its altered character exemplified by its low hydrogen index (200/300), contrasting with the organic matter in the surrounding pelagic/hemipelagic sediments (500/600) but very similar to the organic matter encountered in the near coastal environments of the lake (Fig. 15).

An opposite example where organic-rich mass-gravity flows are delivered into oxic deep ocean environment is provided by Cenozoic sediments deposited on the continental margin off the coast of Angola. A 200 m thick sedimentary sequence of organic-rich sediments (average 3% TOC) has been recovered by drilling in the Southern Angola basin adjacent to the Walvis Ridge at site DSDP 530 (Meyer et al. 1984). These sediments which have been deposited since the Late Miocene were redeposited by mass flows processes, essentially debris flows, originating from shallower water setting on the continental shelf, where organic muds have been previously laid down by pelagic/hemipelagic processes under the highly productive South-West Africa upwelling system, as recognized at site DSDP 532.

If redepositional processes are likely to be significant, for instance in continental slope environments and in tectonically active basins, we can

LAKE TANGANYIKA

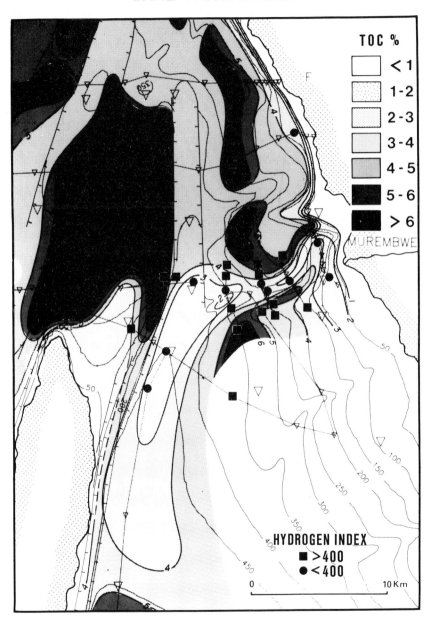

Fig. 15—Distribution of organic matter in a part of Lake Tanganyika.

assume that they should be subordinate in shallow basins with low slope gradients. In such a situation, pelagic and hemipelagic processes are expected to prevail and will result in organic facies distribution basically governed by hydraulic energy generated by waves, currents or storms. According to the previously quoted hydraulic equivalent effect, the organic matter will tend to accumulate in the most quiescent parts of the basin. The belt-like distribution of organic facies which is observed in many present-day anoxic shallow seas and lakes, as exemplified by the Caspian Sea in USSR (Pakhomova 1961) and the lake Bogoria in Kenya (Huc, Herbin et al., submitted), probably corresponds to zonal distribution of hydraulic energy (Fig. 16): the progressive decrease in energy in response to increasing water depth results in a basinward increase in organic concentration which tends to parallel the bathymetric contours. This behavior very often leads to a concentric pattern of the organic concentration in sedimentary basins exhibiting a definite organic increase toward depocenters of lowest energy, in the deeper parts of the basins. This concept may explain the organic distribution observed in certain ancient sedimentary basins of petroleum interest such as the lower Toarcian shales (Fig. 17, after Espitalie et al. 1986).

ORGANIC-RICH SEDIMENTS THROUGH GEOLOGIC TIME

An increasing number of observations support the idea that strata of organic-rich sediments are not randomly distributed in the stratigraphic record, but rather are concentrated in temporarily restricted periods. These intervals include the organic-rich shales of the Devonian, Silurian, Lias, late Jurassic, the largely documented mid-Cretaceous black shales and the major coal deposits during middle and late Carboniferous, middle and upper Permian, lower and middle Jurassic, lower Cretaceous, and late Cretaceous to Eocene (Tissot 1979; Ronov et al. 1980).

Available evidence seems to indicate that the time-specific organic facies are of worldwide extent. Such a situation is obviously a consequence of the relationship between the factors controlling the sedimentation of organic matter and changeable global conditions such as transgressions and regressions, continents configuration, atmosphere and ocean history, and climate. However, if we consider the controversial issues currently debated in the literature (such as (a) mid-Cretaceous black shales are the result of a very high organic productivity, versus (b) mid-Cretaceous black shales have been deposited in a context of low productive sluggish ocean) it becomes obvious that this relationship is imperfectly understood.

Because the organic matter is a geochemical species which cannot be independent at the world scale of the other geochemical species which are involved in the organic cycle (CO_2, SO_4...), we can infer that the understanding of the secular deposition of the organic-rich sediments can be substan-

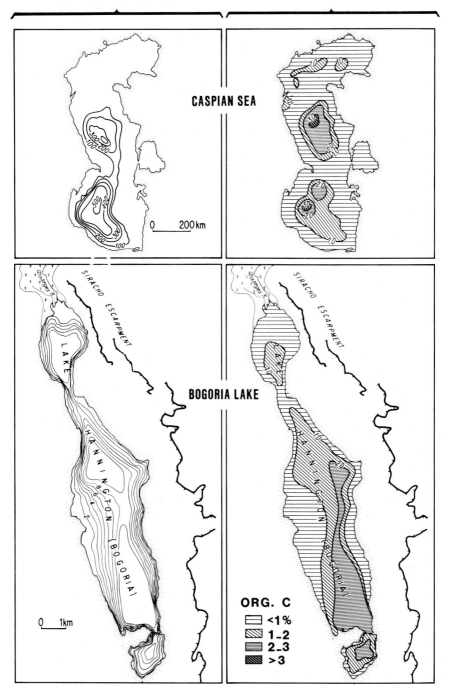

WATER DEPTH (m)   DISTRIBUTION OF ORG. C

CASPIAN SEA

0   200 km

BOGORIA LAKE

0   1km

ORG. C

<1 %
1-2
2-3
>3

Fig. 16—Relationship between topography and organic content in the Caspian Sea (USSR) and in Bogoria Lake (Kenya).

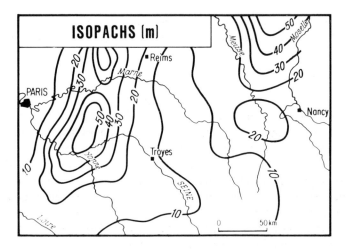

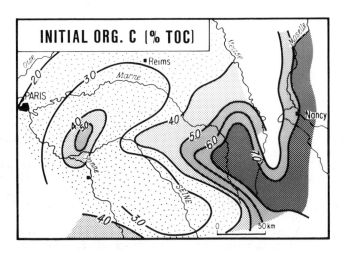

(After Espitalié et al,1986)

Fig. 17—Distribution of initial organic carbon in bituminous shales of lower Toarcian age in the Paris basin.

tially improved with the help of comprehensive models such as those developed by Berner et al. (1983) and Tardy (1986), which integrate the fate of intefering geochemical species into the framework of the global Earth history. However, one should be aware that to be satisfactory these models, which are based on mass balance calculations, have to be constrained, for example by reliable information on the global mass status of the various important types of sediments. This calibration should thus rely on a global inventory of sediments through time. Such an inventory has been attempted in the past years (i.e., Ronov et al. 1980) but is far from complete.

CONCLUSION

Organic sedimentology is a field of research which is still in its infancy. Substantial progress in the understanding of the factors which are involved in the fate of organic matter has been achieved in the last years. The level of primary productivity, nature of biological precursors, extent of biodegradation, and sedimentary processes have been recognized and studied as the basic ingredients controlling the amount, quality and areal distribution of organic matter in sedimentary rocks. However, a better quantitative assessment of these interrelated phenomena is needed. Further development along this direction has to be encouraged in order to change a descriptive/ explanatory discipline into a predictive tool. One of the consequences of such an effort would be to improve our understanding of the uneven distribution of prolific source rocks among geological periods, a distribution which remains a major challenge for sedimentologists and organic geochemists.

REFERENCES

Arthur, M.A.; Dean, W.E.; and Stow, D.A.V. 1984. Models for the deposition of Mesozoic-Cenozoic fine-grained organic-carbon-rich sediments in the deep-sea. In: Fine-grained Sediments, Deep-water Processes and Facies: Geological Society of London, spec. publ. 15: 527–559.

Barron, E.J. 1985. Numerical climate modelling, a frontier in petroleum source rock prediction: results based on Cretaceous simulations. *Bull. AAPG* **69**: 448–459.

Berner, R.A.; Lasaga, A.; and Garrels, R.M. 1983. The carbonate-silicate geochemistry cycle and its effect on atmospheric carbon dioxide over the past 100 million years. *Am. J. Sci.* **283**: 641–683.

Dagg, M.J., and Walser, W.E. Jr. 1986. The effect of food concentration on fecal pellet size in marine copepods. *Limnol. Oceanogr.* **31, 5**: 1066–1071.

Dean, W.E., and Gardner, J.V. 1982. Origin and geochemistry of redox cycles of jurassic to eocene age, Cape Verde basin (DSDP Site 367), continental margin of northwest Africa. In: Nature and Origin of Cretaceous Organic-rich Facies, eds. S.O. Schlanger and M.B. Cita, pp. 55–78. London: Academic Press.

Debyser, J., and Deroo, G. 1969. Faits d'observation sur la genèse du pétrole: facteurs contrôlant la répartition de la matière organique dans les sédiments. *Revue IFP* **24, 1**: 21–48

Demaison, G.J., and Moore, G.T. 1980. Anoxic environments and oil source bed genesis. *Bull. Am. Assn. Pet. Geol.* **64**: 1179-1209.

Degens, E.T., and Mopper, K. 1976. Factors controlling the distribution and early diagenesis of organic materials in marine sediments. In: Chemical Oceanography, 2nd. ed, eds. J.P. Riley and R. Chester. London: Academic Press 6: 60–114.

Degens, E.T.; Von Herzen, R.P.; and Wong, H.K. 1971. Lake Tanganyika: water chemistry, sediments, geological structure. *Naturwissenschaft* **58**: 229–241.

Durand, B. 1983. Present trends in organic geochemistry in research on migration of hydrocarbons. In: Advances in Organic Geochemistry, eds. M. Bjoroy, P. Albrecht, C. Corning, K. de Groot, G. Eglinton, and G. Speers, pp. 117-128. Chichester: John Wiley.

Espitalie, J. 1986. Organic geochemistry of the Paris Basin, 3rd Conferencee on petroleum geology of NW Europe, October 26–29, 1986, London.

Herbin, J.P.; Magniez-Jannin, F.; and Muller, C. 1986. Mesozoic organic rich sediments in the South Atlantic: distribution in time and space. In: Biogeochemistry of Black Shales; Mitt. Geol. Palaont, Inst. Univ. Hamburg, 60.

Huc, A.Y.; Vandenbroucke, M.; Bessereau, G.; and Le Fournier, J.F. 1987. Distribution of organic facies in recent sediments from the northern part of the lake Tanganyika. AAPG meeting, Los Angeles, June 1987.

Ibach, L.E.J. 1982. Relationship between sedimentation rate and total organic carbon content in ancient marine sediments. *Bull. Am. Assn. Pet. Geol.* **66**: 170–188.

Lieth, H., and Whittaker, R.H. 1975. Primary Productivity of the Biosphere. New York: Springer Verlag.

Lisitzin, A.P. 1972. Sedimentation in the world ocean. S.E.P.M. spec. publ. **17**: 218.

Meyer, P.A.; Brassel, S.C.; and Huc, A.Y. 1984. Geochemistry of organic carbon in South Atlantic sediments from Deep Sea Drilling Project Leg 75. In: Init. Repts. DSDP, 75, eds. W.W. Hay, J.C. Sibuet et al. Washington, D.C.: U.S. Govt. Print. Off.

Muller, P.J., and Suess, E. 1979. Productivity, sedimentation rate, and sedimentary organic content in the oceans, 1. Organic carbon preservation. *Deep Sea Res.* **26A**: 1347–1362.

Nwachukwu, J.I., and Baker C., 1985. Variation in kerogen densities of sediments from the orinoco delta, Venezuela. *Chem. Geol.* **51**: 193–198.

Pakhomova, A.C. 1961. Organic carbon in recent sediments from Caspian Sea (in Russian). Cited in: Bordovskiy, O.K. 1969. Organic matter of recent sediments of the Caspian Sea. *Oceanology* **9, 6:** 799–807.

Parrish, J.T. 1982. Upwelling and petroleum source beds with reference to the paleozoic. *Bull. Am. Assn. Pet. Geol.* **66:** 750–774.

Parrish, J.T., and Curtis, R.L., 1982. Atmospheric circulation, upwelling, and organic-rich rocks in the Mesozoic and Cenozoic eras. *Paleogeog. Paleoclim. Paleoecol.* **40**: 31–66.

Parrish, J.T.; Ziegler, A.M.; and Scotese, C.R. 1982. Rainfall patterns and the distribution of coals and evaporites in the Mesozoic and Cenozoic. *Paleogeog. Paleoclim. Paleoecol.* **40**: 67–101.

Pelet, R. 1983. Preservation and alteration of present-day sedimentary organic matter. In: Advances in Geochemistry 1981, ed. M. Bjoroy, pp. 241–250. Chichester: John Wiley & Sons.

Pelet, R. 1984. A model for the biological degradation of recent organic matter. *Org. Geochem.* **6**: 317–325.

Pelet, R. 1987. A model of organic sedimentation on present-day continental margins. In: Marine petroleum source-rocks, eds. J. Brooks and A.J. Fleet. Geological Society Special Publication. 26: 167–180.

Pratt, L.M. 1984. Influence of paleoenvironment factors on the preservation of organic matter in Middle Cretaceous Greenhorn Formation, Pueblo, Colorado. *Bull. Am. Assn. Pet. Geol.* **68**: 1146–1159.

Romankevitch, E.A. 1977. In: Geochemistry of Organic Matter in Oceans (in Russian). USSR: Akad. Nauk.

Ronov, A.B.; Khain, V.E.; Balukhovsky, A.N.; and Seslavinsky, K.B. 1980. Quantitative analysis of Phanerozoic sedimentation. *Sediment. Geol.* **25**: 311–325.

Savdra, C.E.; Bottjer, D.J.; and Gorsline, D.S. 1984. Development of a comprehensive, oxygen-deficient marine biofacies: evidence from Santa Monica, San Pedro, and Santa Barbara basins, California continental borderland. *Bull. Am. Assn. Pet. Geol.* **68**: 1179–1192.

Scotese, C.R., and Summerhayes, C.P. 1986. Computer model of paleoclimate predicts coastal upwelling in the Mesozoic and Cenozoic. *Geobyte* **1, 3**: 28–43.

Shimkus, K.M., and Trimonis, E.S. 1974. Modern sedimentation in Black Sea. In The Black Sea—Geology, Chemistry and Biology, eds. E.T. Degens and D.A. Ross. AAPG memoir 20.

Stanley, D.J. 1986. Turbidity current transport of organic-rich sediments: alpine and mediterranean examples. *Marine Geol.* **70**: 85–101.

Stein. R. 1986. Organic carbon and sedimentation rate. Further evidence for anoxic deep-water conditions in the Cenomanian/Turonian Atlantic ocean. *Marine Geol.* **72**: 199–209.

Suess, E. 1980. Particulate organic carbon flux in the oceans — surface productivity and oxygen utilization. *Nature* **288**: 260–263.

Tardy, Y. 1986. Le cycle de l'eau: climats, paléoclimats et géochimie globale, p. 338. Paris: Masson.

Ten Haven, H.L. 1986. Organic and inorganic geochemical aspects of mediterranean late quaternary sapropels and messinian evaporitic deposits. Thesis, Utrecht University.

Tissot, B. 1979. Effect on prolific petroleum source rocks and major coal deposits caused by sea-level changes. *Nature* **277**: 463–465.

Standing, left to right:
Ed Tipping, Fritz Frimmel, László Hargitai, Richard Zepp

Seated (center), left to right:
Jürgen Niemeyer, Helene Horth, Eva-Christine Hennes, Arie Nissenbaum

Seated (front), left to right:
Jim Weber, Alain Huc, Ulrich Müller-Wegener
(Not shown) Ivan Sekoulov

Humic Substances and Their Role in the Environment
eds. F.H. Frimmel and R.F. Christman, pp. 245–256
John Wiley & Sons Limited
© S. Bernhard, Dahlem Konferenzen, 1988.

Environmental Reactions and Functions Group Report

H. Horth, Rapporteur
F.H. Frimmel
L. Hargitai
E.-C. Hennes
A.Y. Huc
U. Müller-Wegener
J. Niemeyer

A. Nissenbaum
I. Sekoulov
E. Tipping (Moderator)
J.H. Weber
R.G. Zepp

> *Humic substances may not be beautiful, but they do beautiful things.*
> —Fritz Frimmel

INTRODUCTION

In our discussions we tried to focus on the importance of humic substances in the environment and to examine their relevance to environmental processes.

Humic substances comprise a significant proportion of total organic carbon in the global carbon cycle; they are ubiquitous in the environment—in soils, in sediments, in fresh water, and in seawater. They may even be of some importance in the atmosphere, for instance as part of particulate matter in sprays or in rainwater, although little is known about this aspect at present.

The study of the environmental effects of humic substances is particularly difficult because we still have problems in defining humic substances (see Ertel et al., this volume) nor can we determine their chemical identity (see Bracewell et al., this volume). Humic substances are highly complex and, although refractory in character, they have the capacity for diverse chemical and physical interactions in the environment.

Because of the ubiquitous nature of humic substances, and the diversity of their involvement in environmental processes, it was impossible to cover all aspects of environmental reactions and functions. The topics which we covered necessarily reflect to some extent the background and interests of the members of our group.

It was recognized that humic substances in the environment are almost certainly different from isolated humic substances which are operationally defined (see Thurman et al., this volume). Thus the term humic substances in our report refers to these, as present in the environment, unless specified as "isolated humic substances."

EFFECTS OF HUMIC SUBSTANCES ON THE BIOAVAILABILITY OF ELEMENTS AND CHEMICAL COMPOUNDS IN THE ENVIRONMENT

The following topics were identified as being of importance in the discussion:

1) the release of carbon, nitrogen, phosphorus, sulfur, and of metals as a source of nutrients for biota due to degradation of humic substances;
2) the beneficial or deleterious effects on biota caused by binding of metals and organic compounds to humic substances.

As the degradation of humic substances is slow, they do not appear to be a major direct source of the nutrients (carbon, nitrogen, phosphorous, and sulfur) for biota in soil, except for some highly specialized organisms, mainly fungi, which can utilize humic substances as a source of energy. However, the release of the above nutrients may be more important in aquatic systems where humic substances are likely to be transformed more rapidly, e.g., due to photoalteration.

In the context of the possible transport of xenobiotics into cells, the question of penetration of humic substances into cells was considered. Cell penetration by humic substances was indicated in experiments where humic acids could be detected within minutes in vacuoles of plant cells brought into contact with media containing isolated humic acids. Experiments with ^{14}C-labelled, simulated "humic substances" showed ^{14}C-uptake in plant roots, but only minor translocation to the stem. These experiments suggest that xenobiotics which are bound to humic substances may be transported into plant cells, but the problem does not appear to have been addressed sufficiently.

The more mobile, i.e., soluble fractions of humic substances in soil are important extracellular transport media for nutrients; for example, the bioavailability of essential micronutrients (e.g., trace elements) may be increased due to chelation with soluble humic substances and subsequent release. However, chelation of trace elements with the insoluble humic fraction may restrict their bioavailability. Similar processes apply to those metals which can be relevant from a toxicological point of view and to xenobiotics. Binding of these to humic substances and transport may result in increased bioavailability and, therefore, in increased toxicity. Alternatively, binding to the insoluble fraction of humic substances may "immobilize" these and thereby protect biota from deleterious effects. Toxicity is, of course, concentration-dependent; i.e., some metals are essential trace elements at low levels and toxic at higher concentrations. Therefore, the correct balance between availability and "immobilization" is of major importance.

A complex array of different mechanisms is responsible for the differing effects. The effects of humic substances on the bioavailability of toxic organic xenobiotics can often be related to the extent of partitioning between water and humic substances (see below). Factors affecting the bioavailability of metals include oxidation state, solubility, concentration and competition between metals and protons, and complexation by different ligands. Humic substances can act as oxidizing or reducing agents, depending on environmental conditions involving photoprocesses, thereby altering the chemical properties of species associated with humic substances.

Recommended reading

For further information on this broad subject, the reader is referred to Vaughan and Malcolm (1985) and Visser (1986).

Research suggestions

1) Study the effects of humic substance transformation (e.g., photoalteration) on the bioavailability of its carbon, nitrogen, phosphorus, sulfur, and metals.
2) Study the physiological effects of uptake of humic substances by cells and the influence of humic substances on uptake of chemicals by biota.

HOW DO HUMIC SUBSTANCES AFFECT PHOTOCHEMICAL PROCESSES IN WATER AND WHAT ARE THE EFFECTS OF LIGHT ON THE FORMATION AND DEGRADATION OF HUMIC SUBSTANCES?

Recent research has shown that absorption of sunlight by humic substances results in their photolysis as well as in photoreactions of xenobiotics and other organics that occur in water bodies. Direct photolysis of humic substances is poorly understood, but it results in bleaching of the color and formation of small molecules such as carbon monoxide, carbon dioxide, acetone, and acetaldehyde. These reactions are potentially of great importance in carbon cycling in aquatic environments.

Excited states of the humic substances, as well as singlet oxygen derived from electronic energy transfer, have been shown to play an important role in these photoreactions (Braun 1986). By studying photooxidations of furans and xenobiotics it has been found that singlet oxygen formation is induced by solar radiation and that the reactions are surprisingly insensitive to changes in pH, in the pH 5–9 range. These photoreactions also involve superoxide, organoperoxyl, and hydroxyl radicals. Superoxide, through its dismutation, forms hydrogen peroxide, a widely distributed oxidant in the photic zones of freshwaters and the sea. Continuous irradiation in sunlight, as well as laser flash photolysis studies, have shown that hydrated electrons are formed with low quantum yields. Hydrated electrons may be significant in environmental reduction of persistent xenobiotics such as chlorinated pollutants. Photoreactions of insoluble manganese and iron oxides with humic substances result in reductive dissolution and enhanced bioavailability of the metals.

Recommended reading

For an up-to-date account of photochemistry the reader is referred to Zafiriou et al. (1984) and Zika and Cooper (1987).

Research suggestions

Further research needs are in the following areas:

1) Take photochemistry into account in other humic studies (e.g., metal binding).
2) Identify and quantify the photochemically induced reactive species relevant under environmental conditions.
3) Quantification of photoalteration of xenobiotics involving humic substances.
4) Study the role of photoreactions in the carbon cycling of the environment.

THE FORMATION OF CHLORINATED ORGANIC COMPOUNDS FROM HUMIC SUBSTANCES

Humic substances, in water that is abstracted for drinking water treatment, are altered by water treatment processes and are subsequently discharged into the environment in the altered form with effluents.

Much attention has been focused in recent years on the reaction of chlorine with organic matter in drinking water, sewage effluent, and industrial cooling water disinfection. Several hundred chlorinated organic compounds have been identified in drinking waters, yet many more are still unidentified. The presence of chlorinated organic compounds first gave rise to concern with the discovery in the early seventies of chloroform in drinking water (Rook 1974). Although this and other chlorinated organics are present in drinking water at low concentration, the effects of long-term consumption of these and the possible carcinogenic hazard need to be examined. Short-term bacterial assays (e.g., Ames tests) have been employed in many laboratories throughout the world to investigate the mutagenic effects of concentrated extracts of the organic compounds in drinking water. Mutagenic activity was detected in many extracts of drinking water and it has been demonstrated that the mutagenic compounds are produced as a result of the reaction of chlorine with the organics in treated water, with aquatic humic substances, with isolated humic and fulvic acids, and with amino acids.

Only a very small proportion of the chlorination byproducts are mutagenic. Approximately thirty mutagenic compounds have so far been identified, but these account for less than ten per cent of the total mutagenic activity of extracts of drinking water. More recently a chlorofuranone, 3-chloro-4(dichloromethyl)-5-hydroxy-2(5H)-furanone (MX), a highly potent mutagen

first identified in chlorinated pulp effluents, has also been detected in drinking water and in chlorinated humic substances (Hemming et al. 1986). This compound alone appears to account for a significant proportion of the mutagenic activity in extracts of chlorinated humic substances or drinking water. Other similar compounds have also been detected.

Little is known at present concerning the effect of such bioactive compounds on human health, nor of their environmental effect.

A large proportion of the humic substances is still present after chlorination in the form of relatively large molecular weight compounds containing chlorine. This fraction of the chlorinated humic substances needs to be characterized and its biological effects and possible persistence in the environment will need to be examined.

Recommended reading

For a recent review of the subject, the reader is referred to Fielding and Horth (1986). A wide range of recent research reports has been published by Jolley et al. (1985).

Research suggestions

1) To assess the toxicological significance of MX in drinking water and in the environment.

2) To characterize the large molecular weight fraction of chlorinated humic substances and investigate its biological effect and fate in the environment.

THE NATURE OF INTERACTIONS BETWEEN METALS AND HUMIC SUBSTANCES

The following topics were considered:

1) experimental measurement of the extent of humic substance-metal binding;
2) treatment of the data;
3) the nature of metal binding sites; and
4) application of experimental data to natural systems.

The experimental speciation techniques include separation and non-separation techniques. Separation often involves the use of dialysis or ultra-filtration membranes. Nonseparation techniques, which are sensitive enough for environmental samples, include fluorescence quenching, electron spin resonance (ESR) for some metals, and differential pulse anodic stripping voltammetry (DPASV). Because of uncertainties in the effectiveness of the techniques, two or more should be applied in any study.

Knowledge of binding sites for metals in humic substances is very sparse. At present, the best technique is ESR, which can be used to identify donor atoms (e.g., O and/or N) that bind metals. Metal ion nuclear magnetic resonance (NMR) has potential for the study of binding.

It is clear at present that much remains unknown about metal–humic substance interactions within the complexity of environmental systems. For example, in many cases it may be difficult to determine whether or not bound metals are soluble and ionic or colloidal in nature. Because humic substances act as a reducing agent (reduction potential ~ 0.7 volts at pH 0), they can effect, for example, reduction of Fe^{3+} to Fe^{2+}. As a result we do not usually know whether bound iron is divalent or trivalent. The environmental mobility might be partly due to binding and reduction of Fe^{3+} by humic substances in the presence of light. Similar binding and reduction processes might account for mobility of iron in soils.

It is important to carry out competition studies of various metal ions (e.g., Al^{3+}, Ca^{2+}, and trace metals) with H^+ for the anionic binding sites of humic substance. Most of these sites are carboxyl groups which form relatively unstable complexes but N- and S-containing sites suitable for more stable complexes are also present. S sites, for example, would preferentially bind soft (Class B) metals such as Hg^{2+} according to the soft and hard acid and base concept of Pearson. When soluble humic substances are titrated with metals such as Cu^{2+}, aggregation and precipitation occur before all carboxyl sites are occupied due to charge neutralization. Some of these effects might result from polyelectrolyte effects, i.e., variations in net humic charge with metal and/or proton loading.

In order to apply the results of experimental studies of metal binding to environmental systems, mathematical descriptions (models) of the metal–humic interactions are necessary, possibly in combination with generalized inorganic speciation models such as GEOCHEM. The metal–humic models are commonly equilibrium models varying in complexity and range of applicability. The simplest models (e.g., two-site Scatchard; see Weber, this volume) have only a few parameters but are restricted in their use to conditions of constant pH, ionic strength, and concentrations of competing cations. Distribution models, in which variations in the apparent stability constant are assumed to be continuous, are similarly restricted. More elaborate models must be used for environments where ionic strength, pH, and competing metal ion concentrations are expected to vary (e.g., in acid soils and waters subjected to variable rainfall).

A promising approach is that of Marinsky (1986), in which the humics are assumed to have a limited number of types of binding site, at which protons and metals compete for binding, and the effect of variations in net humic charge on binding is taken into account. Proton–humic and metal–humic interactions can, in this model, be characterized by intrinsic equilibrium constants, which remain truly constant (at a given temperature) irrespective of the values of other variables.

Validation of predictive modelling of metal interactions in the environment is an important, although inadequately pursued, aspect. A useful, general approach might involve a continuum of experiments, for example, by stating an initial theoretical model, testing the model in controlled laboratory experiments, reformulating the model as new data are obtained, and finally testing the model on environmental samples with the possibility of reformulating the model to incorporate the field data.

Recommended reading

For further reading the following are recommended:
Saar and Weber (1982), Buffle (1984), and Sposito (1986).

Research suggestions

1) Studies of cation competition effects.
2) Studies of the redox chemistry of metals that interact with humic substances.

3) Investigate the application of metal ion NMR to the study of metal–humic substance binding.
4) Carry out measurements on environmental samples.
5) Devise models based on mechanistic information incorporating physical models, e.g., of water movement, when necessary.
6) Evaluation of modelling should include sensitivity analyses of the model output and take into account the effects of natural variations in humic properties.

INTERACTIONS BETWEEN XENOBIOTICS AND HUMIC SUBSTANCES

It is important to understand interactions between xenobiotics and humic substances as these interactions affect the bioavailability of toxic compounds, transport of these through soils, sediments, and aquatic environments, and may also cause chemical alterations of xenobiotics. Solubility–polarity of a xenobiotic is an important first consideration.

Several binding mechanisms were identified, of which several may occur concurrently: electrostatic forces, hydrogen bonding, hydrophobic interaction, and charge transfer (electron donor/acceptor) complexation.

When organic xenobiotic concentrations are very low in a silt-sized soil (or sediment) suspension in water, the extent of partitioning of the xenobiotic between water and soil can be described approximately by an equilbrium partition coefficient K_p:

$$K_p = \frac{C_s}{C_w} \ (l/kg).$$

In this equation C_s is the xenobiotic concentration on the soil (mg/kg) and C_w is its concentration in the aqueous phase (mg/l) of the suspension. With hydrophobic, nonionic xenobiotics the fraction of organic carbon f_{oc}, on the soil has a major influence on K_p. For a given xenobiotic, the K_p value on a series of soils (or sediments) is approximately proportional to f_{oc}, i.e., $K_p = K_{oc}f_{oc}$ where K_{oc} is the partition coefficient normalized to fractional organic carbon content. The K_{oc} values for a series of xenobiotics are, in turn, proportional to their octanol–water partition coefficients (K_{ow}). Very recent studies have shown that the approach to sorption equilibrium can be very slow in the case of extremely hydrophobic xenobiotics. Sorptive retardation of movement of the xenobiotics into and back out of soil pores and/or aggregates apparently is responsible for this effect.

It was debated whether isolated humic substances could usefully be employed to simulate environmental processes. It was agreed that isolated humic substances could provide a useful tool in studying some binding interactions. Useful data have been obtained, for example, in a relationship between binding and the molecular weight of humic fractions and in a relationship as well as between binding and the degree of aromaticity of the humic substances.

Experimental work with a single compound and extracted humic substances from different aquatic systems has shown quite different results for the different humic substances. This illustrates the need to exercise caution when trying to extrapolate results from one aquatic system to another. In addition, it is essential to validate such results with natural samples before drawing conclusions concerning environmental effects.

Information on binding alone is of limited value. Microbial degradation needs to be considered, as well as chemical alteration due to binding to humic substances and effects of photoprocesses. Such processes could alter xenobiotics to increase or decrease their toxicity. In addition, there is a need to consider binding energies and kinetics which affect the release of bound xenobiotics from humic substances. A need to define binding terminology was identified.

Recommended reading

For background information the reader is referred to Goring and Hamaker (1972). An overview of research on sorptive interactions between xenobiotics and soil or aquatic humic substances is provided by Chiou et al. (1986), and a recent review article by Karickhoff (1984) discusses the role played by particle-associated organic matter in controlling xenobiotic sorption. In addition, Führ (1987) is recommended.

Research suggestions

1) Establish whether the partition coefficient is sufficient for modelling binding interactions.
2) In addition to well-established parameters, such as the octanol/water partition coefficient, we need characterization data, including binding data, to study the influence of humic substances on biological effects and chemical reactions of xenobiotics.
3) Modelling of the interactions with a view to environmental prediction should be carried out along the same general lines as for metal binding, perhaps with more emphasis on kinetics.

SORPTION OF HUMIC SUBSTANCES ON SURFACES AND ENVIRONMENTAL CONSEQUENCES

The use of thermodynamic parameters was discussed to describe the energy of adsorption of humic substances on surfaces. This energy is composed of several energy values: chemical, electrostatic, hydrophobic, van der Waals forces, hydrogen bonding, metal/proton interactions, and conformation of humic molecules. It is not possible at present to measure these values, although adsorption measurements may be made under controlled conditions, but the results of such experiments are difficult to extrapolate to the natural environment.

However, it is important to be able to quantify adsorption energy in order to estimate environmental consequences resulting from colloid formation, complexation of humic substances with solid surfaces, transport of humic substances, crystallization processes influenced by humic substances, and changes in physiological properties of humic substances. Adsorption energies are affected by pH, the physical characteristics of the surfaces, the concentration of humic substances, as well as the concentration of dissolved inorganic compounds.

Adsorption of humic substances on surfaces results in fractionation of the humic substances, i.e., selected molecules are adsorbed preferentially and the characteristics of the humic substances remaining in solution are altered as a result.

Organic coating of surfaces dictates further interactions. The cation exchange capacity of clay surfaces, for instance, and their catalytic activity are altered by surface coating as there is the stability of colloidal suspensions. Electron microscopy has been applied to examine surface coating.

The main vehicles of transport for humic substances in the aquatic environment need to be explored. Such studies should include an assessment of the biological packing of organic and mineral surfaces (fecal pellets, wrapping of organic and mineral surfaces by mucopolysaccharides which leads to formation of aggregates) which might result in removal of humic substances and xenobiotics and metals from the hydrosphere by sedimentation.

Fractal dimensions have been applied to the study of mineral surfaces in order to describe the degree of irregularity, not as a deviation from ideal conditions but as an intrinsic value (Pfeifer 1984). The physical properties of surfaces involved in humic substance adsorption vary a great deal and may also be usefully described using fractal dimensions.

Although a considerable volume of adsorption data is available in the literature, more information on natural samples is required and additional information on the physical chemistry of humic substances is needed.

Recommended reading

Theng (1979) is recommended for further information on sorption in soils and Tipping (1986) for a review of sorption in aquatic systems.

Research suggestions

1) Investigate how adsorption of humic substances on environmental surfaces affects the reactivity of adsorbent and adsorbate, including photoprocesses.
2) Investigate the effect of biological activity on the immobilization of metals and xenobiotics in the hydrosphere by sedimentation.
3) Improve the understanding of physical chemistry of humic substances in solution.
4) Explore the application of fractal calculations for characterization of surfaces and sorptive binding sites in humic substances.

CONCLUDING REMARKS

The most frequently recurring problem encountered in our discussions was the lack of experimental data for real environmental samples. Although the need to study simplified model systems in carefully controlled experiments in the laboratory is recognized, the need to validate experimental data on environmental samples cannot be overemphasized.

It is important to consider humic substances as complex systems in the environment and to distinguish between undissolved and dissolved fractions in an exchange process.

Humic substances appear to exert a stabilizing effect on environmental processes, for example by assimilation and subsequent slow release of chemical compounds they act as a reservoir of trace nutrients and contaminants. Chemical partition of chemicals into humic substances results in a "buffering" of the environmental mobility of chemicals. Their function, however, as scavengers of pollutants is limited and can be overtaxed. Care should be taken not to overload the system and to prevent long-term damage which may not be apparent until remedial action becomes difficult and perhaps impossible.

REFERENCES

Braun, A.M.; Frimmel, F.H.; and Hoigné, J. 1986. Singlet oxygen analysis in irradiated surface waters. *Int. J. Env. Analyt. Chem.* **27**: 137–149.

Buffle, J. 1984. Natural organic matter and metal-organic interactions in aquatic systems. In: Metal Ion in Biological Systems, ed. H. Sigel, vol. 18, pp. 166–221. New York: Marcel Dekker.

Chiou, C.T.; Malcolm, R.L.; Brinton, T.I.; and Kile, D.E. 1986. Water solubility enhancement of some organic pollutants and pesticides by dissolved humic and fulvic acids. Env. Sci. Tech. 20: 502–508.

Fielding, M., and Horth, H. 1986. Formation of mutagens and chemicals during water treatment chlorination. Wat. Supply 4: 103–126.

Führ, F. 1987. Non-extractable pesticide residues in soil. In: Pesticide Science and Biotechnology, ed. R. Greenhalgh and F. Roberts. Oxford: Blackwell Scientific Publ.

Goring, C.A.J., and Hamaker, J.W. 1972. Organic Chemicals in the Soil Environment. New York: Marcel Dekker.

Hemming, J.; Holmbom, B.; Reunanen, M.; and Kronberg, L. 1986. Determination of the strong mutagen 3-chloro-4-(dichloromethyl)-5-hydroxy-2(5H)-furanone in chlorinated drinking and humic waters. Chemosphere 15: 549–556.

Jolley, R.L.; Bull, R.J.; Davis, W.P.; Katz, S.; Roberts, M.H. Jr.; and Jacobs, V.A. 1985. Water Chlorination: Chemistry, Environmental Impact and Health Effects, vol.f 5. Chelsea, MI: Lewis Publishers Inc.

Karickhoff, S.W. 1984. Organic pollutant sorption in aquatic systems. J. Hydraul. Eng. 110: 707–735.

Marinsky, J.A., and Ephraim, J. 1986. A unified physicochemical description of the protonation and metal ion complexation equilibria of natural organic acids (humic and fulvic acids). i. Analysis of the influence of polyelectrolyte properties on protonation equilibria in ionic media: fundamental concepts. Env. Sci. Tech. 20: 349–354.

Pfeifer, P. 1984. Fractal dimension as working tool for surface-roughness problems. Appl. Surf. Sci. 18: 146–164.

Rook, J.J. 1974. Formation of haloforms during chlorination of natural waters. Water Treat. Exam. 23: 234–243.

Saar, R.A., and Weber, J.H. 1982. Fulvic acid: modifier of metal-ion chemistry. Env. Sci. Tech. 16: 510A–517A.

Sposito, G. 1986. Sorption of trace metals by humic materials in soils and natural waters. C.R.C. Crit. Rev. Env. Control 16: 193–229.

Theng, B.K.J. 1979. Formation and Properties of Clay-Polymer Complexes. Amsterdam and New York: Elsevier.

Tipping, E. 1986. Some aspects of the interactions between particulate oxides and aquatic humic substances. Marine Chem. 18: 161–169.

Vaughan, D., and Malcolm, R.E. 1985. Soil Organic Matter and Biological Activity. Dordrecht: Martinus Nijhoff.

Visser, S.A. 1986. Effects of humic substances on plant growth. In: Humic Substances, Effect on Soil and Plants. Rome: REDA.

Zafiriou, O.C.; Joussot-Dubien, J.; Zepp, R.G.; and Zika, R.G. 1984. Photochemistry of natural waters. Env. Sci. Tech. 18: 358A–371A.

Zika, R.G., and Cooper, W.J., eds. 1987. Photochemistry of Environmental Aquatic Systems. ACS Symposium Series 327. Washington, D.C.: American Chemical Society.

List of Participants with Fields of Research

G. ABBT-BRAUN
Institut für Wasserchemie und
Chemische Balneologie der
Technischen Universität
Marchioninistr. 17
8000 München 70, F.R. Germany

Derivatization and pyrolysis of aquatic
humic substances

G.R. AIKEN
U.S. Geological Survey—WRD
5293 Ward Road
Arvada, CO 80002, U.S.A.

Analysis of organic acids in water

P.BEHMEL
Institut für Bodenwissenschaften
Abt. Chemie und Biochemie im System
Boden
Universität Göttingen
Von-Siebold-Str. 2
3400 Göttingen, F.R. Germany

Synthesis, derivatization, and analysis
of humic substances; analysis of soil
organic matter, chemistry of model
substances, spectroscopy

J.M. BRACEWELL
Macaulay Land Use Research Institute
Craigiebuckler
Aberdeen AB9 2QJ, Scotland

The nature of soil organic matter as
reflected by analytical pyrolysis, in
relation to its evolution, agronomic
properties and environmental effects

R.F. CHRISTMAN
Dept. of Environmental Sciences and
Engineering, School of Public Health
Room 106B, Rosenau Hall 201H
University of North Carolina
Chapel Hill, NC 27514, U.S.A.

Humic acid chemistry

J.W. DE LEEUW
Dept. of Chemistry and Chemical
Engineering
Organic Geochemistry Unit
University of Delft
de Vries van Heystplantsoen 2
2628 RZ Delft, Netherlands

Organic geochemistry

J.R. ERTEL
Chemistry Dept.
Woods Hole Oceanographic Institution
Woods Hole, MA 02543, U.S.A.

Lipid and carotenioid diagenesis in
sediments; marine DOC
characterization

M. EWALD
Groupe d'Oceanographie Physico-
Chimique
Laboratoire de Chimie Physique A
Université de Bordeaux I
351 Cours de la Liberation
33405 Talence Cedex, France

Physicochemical oceanography

W.R. FISCHER
Lehrstuhl für Bodenkunde
Technische Universität München
8050 Freising-Weihenstephan
F.R. Germany

Heavy metal complexation, properties
of water soluble humic substances,
humus in "underwater soils"

W.J.A. FLAIG
Otto-Hahn-Str. 132
8708 Gerbrunn bei Würzburg.
F.R. Germany

Biochemistry of humic substances

U. FÖRSTNER
Arbeitsbereich Umweltschutztechnik
Technische Universität Hamburg-
Harburg
Eissendorfer Str. 38
2100 Hamburg 90, F.R. Germany

Environmental engineering,
geochemistry

F.H. FRIMMEL
Engler-Bunte-Institut der Universität
Karlsruhe
Richard-Willstätter Allee 5
7500 Karlsruhe 1, F.R. Germany

Chemical reactions in aquatic systems
and in water treatment

A.H. HACK
Institut für Hygiene
Ruhr Universität Bochum
Universitätsstr. 150
4630 Bochum 1, F.R. Germany

The influence of humic substances on
bacterial activity in groundwater

K.M. HAIDER
Institut für Pflanzenernährung und
Bodenkunde (FAL)
Bundesallee 50
3300 Braunschweig, F.R. Germany

Biochemistry of formation and
degradation of humic substances

L. HARGITAI
Dept. of Soil Science
University of Horticulture
Villányi út 35-43
1118 Budapest, Hungary

Chemistry and biochemistry of humic
substances, their role in the
environment

G.R. HARVEY
NOAA/Atlantic Oceanographic and
Meteorological Laboratory
4301 Rickenbacker Causeway
Miami, FL 33149, U.S.A.

Marine humic substances and marine
atmospheric chemistry

P.G. HATCHER
U.S. Geological Survey
923 National Center
Reston, VA 22092, U.S.A.

Geochemistry (organic) in the origin of
coal and humic substances

M.H.B. HAYES
Chemistry Dept.
University of Birmingham
Edgbaston, Birmingham B15 2TT,
England

Structures of humic substances,
interactions of humic substances with
organic chemicals

J.I. HEDGES
School of Oceanography, WB-10
University of Washington
Seattle, WA 98195, U.S.A.

Geochemistry of the major
biochemicals and humic substances in
aquatic environments

E.-C. HENNES
Institut für Radiochemie, Abt.
Wassertechnologie
Kernforschungszentrum Karlsruhe
Postfach 3640
7500 Karlsruhe 1, F.R. Germany

Chemical speciation in aquatic systems, their modeling with computer-aided programs; influence of anthropogenic pollutants on processes in underground with regard to groundwater quality

H. HORTH
Water Research Centre, Medmenham
Laboratory
Henley Road, Medmenham,
P.O. Box 16
Marlow, Bucks. SL7 2HD, England

Identification of mutagens which are produced from the reaction of chlorine with naturally occurring organics in water

A.Y. HUC
Institut Français du Pétrole, B.P. 311
92506 Rueil Malmaison Cedex, France

Organic sedimentology

R.F.C. MANTOURA
Institute for Marine Environmental
Research
Prospect Place, The Hoe
Plymouth PL1 3DH, England

Marine biogeochemistry

J.P. MARTIN
Dept. of Soil and Environmental
Sciences
University of California
Riverside, CA 92521, U.S.A.

Genesis of soil humus; properties of humic acid-type fungal polymers

U. MÜLLER-WEGENER
Institut für Bodenwissenschaften
Abt. Chemie und Biochemie im System
Boden
Universität Göttingen
Von-Siebold-Str. 2
3400 Göttingen, F.R. Germany

Soil chemistry and biochemistry

J. NIEMEYER
Institut für Bodenwissenschaften
Abt. Chemie und Biochemie im System
Boden
Universität Göttingen
Von-Siebold-Str. 2
3400 Göttingen, F.R. Germany

Sorption phenomena in soils; soil surface chemistry

H.H. NIMZ
Institut für Holzchemie
Leuschnerstr. 91
2050 Hamburg 80, F.R. Germany

Wood chemistry

A. NISSENBAUM
Academic Secretary's Office
Weizmann Institute of Science
Rehovot 76100, Israel

Biogeochemistry and organic geochemistry

D.L. NORWOOD
Analytical and Chemical Sciences
Research Triangle Institute
P.O. Box 12194
Research Triangle Park, NC 27709, U.S.A.

Water chemistry

J.W. PARSONS
Dept. of Soil Science
University of Aberdeen
Meston Walk
Aberdeen AB9 2UE, Scotland

Chemistry of soil organic matter

E.M. PERDUE
School of Geophysical Sciences
Georgia Institute of Technology
Atlanta, GA 30032, U.S.A.

Equilibrium chemistry of humic substances, especially proton and metal binding; structural chemistry of humic substances

F.K. PFAENDER
Dept. of Environmental Sciences and
Engineering
University of North Carolina
Chapel Hill, NC 27514, U.S.A.

Microbial transformations of organic
compounds in the environment

R. POCKLINGTON
Bedford Institute of Oceanography
P.O. Box 1006
Dartmouth, Nova Scotia B2Y 4A2,
Canada

Marine dissolved, particulate and
sedimentary organic matter.

M. SCHNITZER
Land Resource Research Center
Agriculture Canada, Central
Experimental Farm
Ottawa, Ontario K1A 0C6, Canada

Chemistry of humic substances;
chemistry of the "unknown" soil
nitrogen

H.-R. SCHULTEN
Dept. of Trace Analysis
Fachhochschule Fresenius
Dambachtal 20
6200 Wiesbaden, F.R. Germany

Thermal degradation of
macromolecules; pattern recognition;
structural chemistry of
humic substances

I. SEKOULOV
Technische Universität Hamburg-
Harburg
Eissendorfer Str. 38
2100 Hamburg 90, F.R. Germany

Advanced wastewater treatment
(nitrification, denitrification, removal of
refractory CSB, phosphorus, suspended
solids, etc.)

F.J. STEVENSON
Dept. of Agronomy, Turner Hall
University of Illinois
1102 S. Goodwin Avenue
Urbana, IL 61801, U.S.A.

Chemistry of soil humic substances:
emphasis on isolation and extraction,
reactive functional groups, and metal-
ion binding

R.S. SWIFT
Dept. of Soil Science
Lincoln College
Canterbury, New Zealand

Isolation, fractionation, and physical
chemistry of soil humic substances and
their interaction with inorganic soil
components

B. SZPAKOWSKA
Dept. of Agrobiology and Forestry
Swierczewskiego 19
60-809 Poznań, Poland

Isolation of humic substances dissolved
in water; properties of dissolved humic
substances

E.M. THURMAN
U.S. Geological Survey, WRD
P.O. Box 25046, Mail Stop 418
Denver Federal Center
Lakewood, CO 80225, U.S.A.

Organic geochemistry of groundwater;
chemistry of dissolved organic
substances (humic substances)

E. TIPPING
Freshwater Biological Association
The Ferry House
Ambleside, Cumbria LA22 0LP,
England

Environmental surface and colloid
chemistry

S.A. VISSER
Dept. of Soil Science (Agriculture)
Laval University
Ste-Foy
Québec, G1K 7P4, Canada

Physiological effects of humic
substances on microorganisms and on
higher plants

J.H. WEBER
Chemistry Dept., Parsons Hall
University of New Hampshire
Durham, NH 03824, U.S.A.

Humic matter-metals interactions;
environmental processes of
organometallic compounds

M.A. WILSON
CSIRO, Division of Fossil Fuels
P.O. Box 136
North Ryde, NSW 2113, Australia

Applications of NMR spectroscopy in
organic geochemistry

R.G. ZEPP
U.S. Environmental Protection Agency
Environmental Research Laboratory
College Station Road
Athens, GA 30613, U.S.A.

Environmental chemistry; kinetics of
sunlight-induced and oxidative
processes in water and on
environmental surfaces

W. ZIECHMANN
Interfakultatives Lehrgebiet Chemie
Universität Göttingen
Von-Siebold-Str. 2
3400 Göttingen, F.R. Germany

Chemical systems in soils

Subject Index

263

Author Index